My Magnificent Machine

William L. Coleman

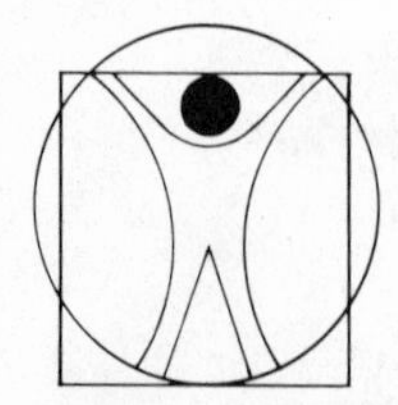

My Magnificent Machine

Published by Bethany Fellowship, Inc.
6820 Auto Club Road, Minneapolis, Minnesota 55438

Printed in the United States of America

Library of Congress Cataloging in Publication Data

Coleman, William L.
My magnificent machine.

SUMMARY: An explanation of basic human physiology interspersed with Biblical references and related religious thoughts and prayers.

1. Human physiology—Juvenile literature. 2. Body, Human—Juvenile literature. 3. Children—Religious life—Juvenile literature. [1. Body, Human. 2. Christian life]. I. Title.
QP37.C693 612 78-5035
ISBN 0-87123-381-9

Dedicated to

The Sterling Evangelical Mennonite Church
Sterling, Kansas

Acknowledgement

to Jim Coleman for his technical assistance

Biographical Sketch

William L. Coleman is a graduate of the Washington Bible College in Washington, D.C., and Grace Theological Seminary in Winona Lake, Indiana.

He has pastored three churches: a Baptist church in Michigan, a Mennonite church in Kansas and an Evangelical Free Church in Aurora, Nebraska. He is a Staley Foundation lecturer.

The author of 75 magazine articles, his by-line has appeared in *Christianity Today*, *Eternity*, *Good News Broadcaster*, *Campus Life*, *Moody Monthly*, *Evangelical Beacon*, *The Christian Reader*.

Presently Mr. Coleman devotes his time as a professional writer.

Bill and his wife, Pat, are the parents of three children. They make their home in Aurora, Nebraska.

Contents

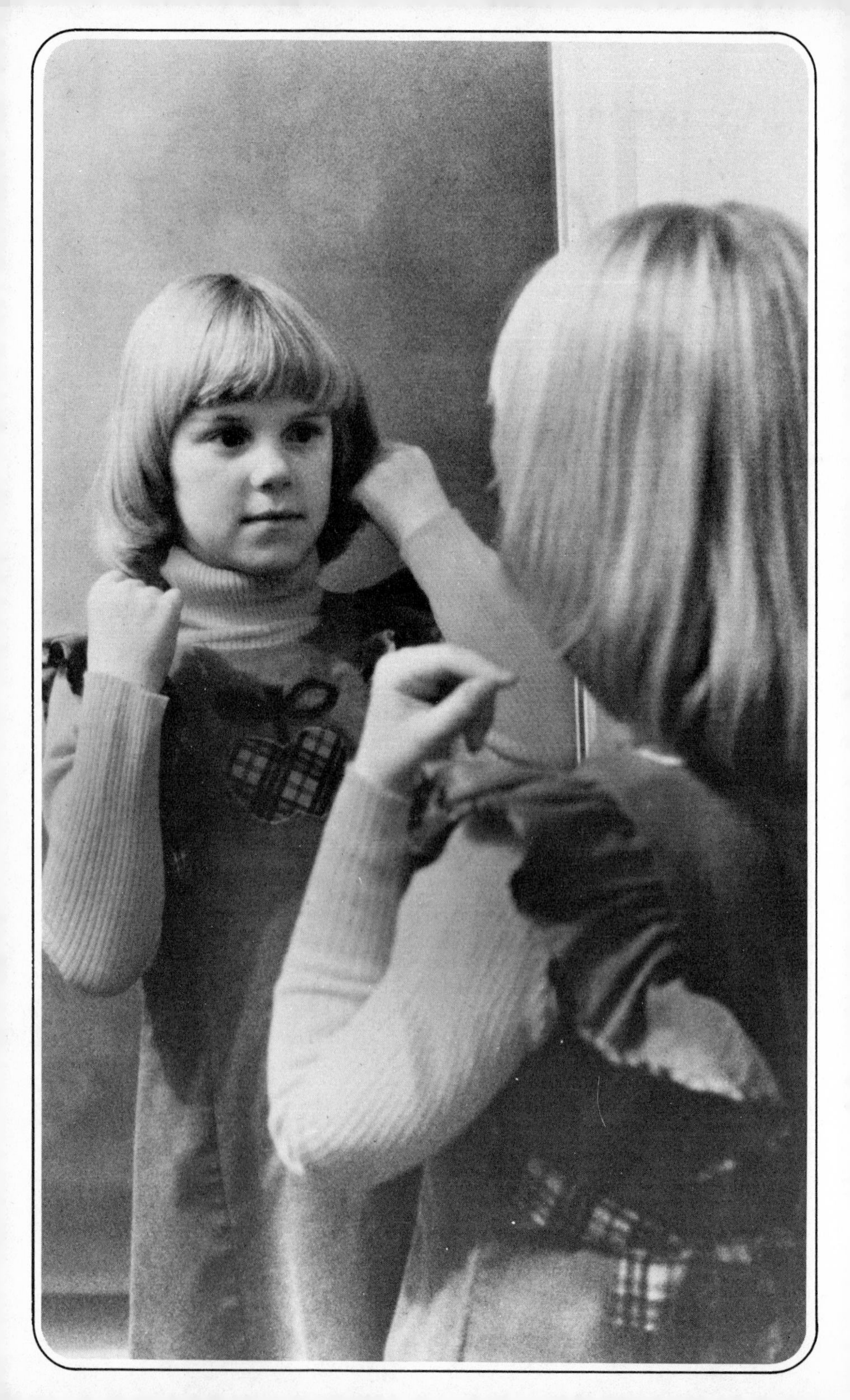

Just Look At Yourself

What kind of body do you have? It has 60,000 miles of tubing just to carry blood. Every time you catch a ball, billions of cells go to work and 200 muscles have to move in a split second.

Messages travel to our brain like a rocket at 350 feet per second. Yet for all our moving parts no one else in all history is like you. No one even smells the way you do.

We have a careful Creator. When we look at ourselves, we will understand him better.

Read and think about what God has done. He is the same God who sent Jesus Christ to earth.

William L. Coleman
Memorial Day, 1977
Aurora, Nebraska

one

The Way I Am Made

It would be exciting to visit the Himalayan Mountains of Nepal or take a ride on an Explorer across the ocean floor. However, no adventure is guaranteed to be more fascinating than taking an unbelievable look at our own bodies. A body with more nerves and cells and more harmony than science can hope to match.

Take just one part of our system we seldom think of—our liver. A simple 3-pound organ, the liver is called upon to carry out over 500 jobs in a person. Hopefully without a hitch.

The liver is located beneath our ribs and stands guard over the chemicals which come into our body. If it were not for the liver, coffee, nicotine and alcohol would kill us. Even chocolate candy would be bad for us. If our liver ever stops working, we will each be dead in 8 to 24 hours.

Not only does it fight off the harmful effects on our system, but also it stores up the good things. Our liver will hold important vitamins such as A, D and B. That is the reason why eating the liver from animals is so good for us. The health-giving iron is also kept available in this organ. Our sugar level is maintained the same way.

Some people abuse their livers by the things they eat and drink, but the liver can take a lot. For instance, if disease destroyed 85% of our liver, we could still go on living and the tough little organ would do the job. It is so capable that given time and good health it wouldn't be too long before the liver cells would grow back the large part it lost.

The liver also makes an excellent medicine chest. Its small factory produces something called globulin which stays busy tackling diseases when they enter the body.

As tough as our liver is, it has to be treated with respect. Too much alcohol can create a problem called cirrhosis and could even ruin the liver. When a person gets too overweight his liver will also get fat and hurt its performance.

Miraculous and helpful as it is, the liver is only one part of an unbelievably efficient machine. If God had put it together in just a slightly different way, man would never be able to exist. However, we can be thankful for a Master Architect who laid it out perfectly.

"You made all the delicate, inner parts of my body, and knit them together in my mother's womb. Thank you for making me so wonderfully complex! It is amazing to think about. Your workmanship is marvelous—and how well I know it" (Ps. 139:13-14).

1. How many jobs does a liver do?
2. Which vitamins are held by the liver?
3. Why did God give us a liver?

Thank you for making me very well.

two

The Magnificent Machine

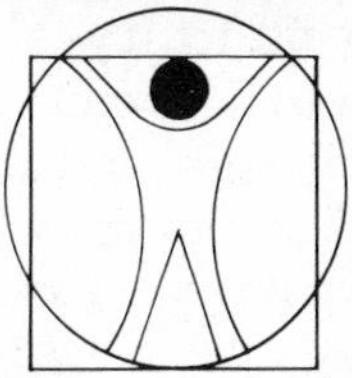

Man doesn't have the biggest brain in the world, just the best. An elephant's brain weighs 11 pounds. A whale's can be 4 1/2 pounds, while a man's adds up to 3 pounds. However, there is no animal which can come close to what a man can think. In fact, there is no computer which has the ability of a human mind. If we could build a machine to do everything a brain does and as quickly, the mechanical brain would be as big as a battleship.

Consider how complicated this system is. The brain has 30 billion nerve cells which in turn connect into other cells. Some of these contact 60,000 other cells. Just to scratch our nose or to catch a ball, billions of these cells have to work at the exact moment in cooperation with all the others. In order for us to stand up, the brain has to send out messages to 200 muscles. Instantly they all have to work together just to get us out of a chair. The brain is exactly set and may run for 70-100 years without a serious breakdown.

The brain is carefully packed in a 6 to 8 inch case called a skull. To protect it from serious bumps and shocks it rests in a watery fluid.

Often we think of the brain as one organ in our body, but in fact it has several sections. Basically it has three major divisions.

The thinking part is called the cerebrum. It takes up about 85% of the brain. While we are reading this chapter the cerebrum is working hard to process the words and thoughts very rapidly. If there is food cooking in the kitchen, the cerebrum is deciding if it smells good or not. If the radio is playing or people are talking, the cerebrum is racing. It picks up a little bit here and a small sound there. When we make decisions, a wish or if we dream tonight, it will again be this large section doing the job.

Different parts of the cerebrum do their own jobs. One little section controls speech, another allows us to hear and yet another controls our memory. Some scientists believe everything we have ever thought or seen is still stored in our brain. We may have difficulty remembering, but the information is still locked in our cerebrum.

The second major division is called the cerebellum. No larger than a rubber ball it gives commands to our muscles. It controls our coordination, allows us to run, jump and dodge.

The third division is the medulla. This section controls our heart, lungs, stomach, blood, breathing and generally keeps our body running smoothly.

Here's an easy way to remember the three parts:

(1) *cerebrum controls thinking*
(2) *cerebellum controls muscles*
(3) *medulla controls body organs*

Because our cerebrum gives us the power to choose, we can control our own brains. If we want to think rotten and cruel things, our brain will do what we want. If, on the other hand, we want to think about helping people, being kind and praising God, the choice is equally ours. God wants us to control our own minds.

"Fix your thoughts on what is true and good and right. Think about things that are pure and lovely, and dwell on the fine, good things in others. Think about all you can praise God for and be glad about" (Phil. 4:8).

1. Name the three parts of the brain.
2. Why does a fluid surround our brain?
3. What does God want us to think about?

Help us to think about bright and good things rather than the gloomy.

three

Funny Medicine

In man's search for good health, he has done some silly things. We have made up medicines and cures which are only funny today.

Some people used to believe plants were shaped like the medicine they could be. If a person had heart trouble, he was fed leaves with a heart shape. If they thought he had worms, he ate wormy looking plants. Did he need hair for his head? How about a hairy looking plant? We can only guess how many people were made sick by these cures.

You were in for a real treat if you believed goat tea was a good medicine. This special drink was reported to cure diseases. Goat tea was made from a mixture of goat, lamb or sheep manure.

The most popular way of keeping illness away was to wear medicine around the neck. If you wanted to keep a sore throat away, you wore a necklace of gold beads. To stop a cold from settling in your chest, you wore a dirty sock around your neck. To keep nosebleeds away, you merely wore a piece of red yarn.

To control sickness all sorts of things were placed against the patient. If someone had the chills, a chicken would be killed and its warm body placed against the sick person. If he had pneumonia, an onion could pull him out. Merely cut one in half and place the middle against the soles of the feet. Some people tried to cure earaches by spitting tobacco juice into the ear.

In no area were there more false stories than in having babies. People believed certain drinking water made it easy to become pregnant. A baby born on a stormy night was supposed to be nervous and cross. A baby born at 4 p.m. would become rich. If a mother gave the baby clothes away too soon, she would have another child.

During the time of Christ the doctors were more limited in their knowledge of medicine. Some of them did excellent work for that day. Others did little more than hurt people. One lady came to Christ after "she had suffered much from many doctors through the years" (Mark 5:26).

A great many stories are told about medicine which are not true. People still tell some of these strange cures. Many people are teaching odd things in the name of religion also. They say a great many things about God which are not true. The best teacher about Christianity is God's Son, Jesus Christ. He understands both God and man and brings us together.

"Jesus told him, 'I am the Way—yes, and the Truth and the Life. No one can get to the Father except by means of me' " (John 14:6).

1. What was goat tea?
2. What were dirty socks used for?
3. What is the best way to find out about God?

People believe many things about the body which are not true. Help us to know the truth about God by knowing Jesus Christ.

four

The Red River

If you could get into a tiny canoe and travel through someone's blood system, where would you go? So we won't miss anything, let's start where the blood begins. We board our boat inside the bones. The center of the bone, called marrow, is the factory which makes the liquid.

Small babies can make blood from all of their bones. But as we grow older, our ribs, backbones and flat bones do most of the work.

After leaving the bones it begins a trip through our body. It can travel up our legs as well as down because of the one-pound pump called heart.

Hold on because this trip is going to be fast. Blood will make it completely around our body and back in less than one minute. And it will do it thousands of times each day.

The trip has a good purpose. Blood is needed to deliver both food and oxygen throughout our body. It also acts as garbage collector and removes waste along the way.

As blood travels, it goes along three highways. They are called veins, arteries and capillaries.

Is our blood really red or are there some real blue bloods? The truth is that all of us have both red and blue blood. When it comes out of our heart in the arteries, the color is red. It has oxygen and hemoglobin. As it travels through the body it gives off the food and oxygen we mentioned. The loss of oxygen makes the blood turn dark and bluish. As it heads back toward the heart (through the veins) it can be called blue blood. The heart turns it red and the blood begins the trip again.

How much blood a person has depends on his weight. If you weigh 100 pounds, your body probably holds around seven pounds of blood.

Not everyone has the same type of blood. There are four regular types. If anyone loses blood in an accident or for other reasons, he can get more. Another person with his type can give him blood. The doctor or nurse removes blood from one person's arm and puts it into the other person.

Type of blood has nothing to do with color of skin or the nation a person comes from. An Indian may have the same kind as a Chinese. A German could have the same as an Egyptian. A black man's may be the same as a white.

Blood is at the center of human life. If it stopped flowing, we would have to die. Jesus Christ is the center of life forever. He made it possible for us to live with him after this life is over.

"I have written this to you who believe in the Son of God so that you may know you have eternal life" (1 John 5:13).

1. When do we have "blue" blood?
2. How does blood help our body?
3. What life does Christ give us?

Christ has given us life even after this body stops working.

five

Bumps and Boils

As long as man has lived he has had trouble with his skin. The sun has burned it, spears have put holes in it and warts grow in the strangest places.

One of the earliest men mentioned in the Bible was a man with a skin problem. His name was Job. He had painful boils on his body.

There probably aren't many people who make it through life without a few boils. They are bumps which rise on the skin for a short time. Sometimes they are ugly. They may even hurt.

Most boils are little pests. However, some are as large as half a dollar and should be taken seriously.

Often a boil begins this way: the skin becomes rubbed and broken. Maybe a shirt collar is too tight and after scratching all day finally breaks through the skin. They can come from picking at one's face or a small cut.

The minute the skin is broken, our body sends an army of doctors to help. Corpuscles come quickly to help fight off any germs which are trying to get into the body. Usually the boil is made up mainly of corpuscles and pus.

Most of us like to break boils to get rid of them. If we do this too soon, we may drive the germs deeper into the skin. We may also get more germs from our fingers. Then more corpuscles are called for and the healing is only slowed.

Some people seem to get a lot of boils. They may have a skin problem, a health difficulty or are not keeping themselves clean enough. The person may not be getting the right food and exercise.

Not all boils stand alone. Sometimes three or six group together in one spot. The bunches of boils are called carbuncles. They are more dangerous because a group can lead to serious infection.

The Bible tells us Satan gave Job a bad case of the boils. They covered him from head to foot. Job was so uncomfortable he sat in a hill of ashes and scraped himself with a piece of broken pottery. He must have felt miserable. The scraping could only add to his problems and infection.

King Hezekiah's health had fallen apart and his body was once covered with boils. He was so sick he was about to die. Finally God agreed to let the ruler live. Then Isaiah told the servants to prepare a medicine to clear up the king's skin.

"For Isaiah had told Hezekiah's servants, 'Make an ointment of figs and spread it over the boil, and he will get well again' " (Isa. 38:21).

1. What causes boils?
2. What is pus?
3. How can a mind fight hate?

Our body is fast in fighting disease and dirt. A mind needs to be just as careful to fight hate, jealousy and meanness.

six

Getting Good Exercise

We often say playing is for children. But we are now learning that children who learn to play well often grow up to be healthy adults.

Many adults and some children are often sick because they have not learned to exercise. Our bodies were made strong for action. The best way to keep them strong is by using them.

The lungs do not hold as much air as they would like. The heart doesn't beat as fully as it might want to.

In recent years young people in North America have taken physical fitness tests. The youth of many other countries do far better than Americans. We may be living a long time but not be as healthy.

Possibly the best exercise program a person can have is to learn gymnastics. Tumbling, trampoline jumping, chin-ups, and the parallel bars help keep a body in excellent condition. This sport is also good because it can be done alone. If you can't find a friend to join you, there is no need to call it off. Started as a child, it is an easy sport to continue.

Running is good exercise to help the heart, lungs and keep the weight down. It also gives strength to our legs. People who haven't done it recently should start off slowly.

Skating will do the same thing for our bodies but in small amounts. It does not help as much as running, but it may be a better way for some to start.

Swimming is good for both the muscles and the breathing. The entire body gets involved. A lot of time spent at the pool is wasted with roughhousing and getting a tan. However, the person who really swims keeps his health sharp and burns up calories.

Tennis is a growing sport. Now there are courts in public parks all over the country. It is a good sport for those of middle age—not as tiring as gymnastics nor as slow as walking. Tennis is good for the arms, legs and increases the breathing. If done correctly it can be played until we are near retirement age.

Riding bikes is a healthy sport and easy to do at your own speed. If the person isn't used to exercising, he can peddle slowly. As he feels stronger he can go faster. It is good for the legs, arms and the breathing if done fast.

As children go through school they probably will play a lot of football, baseball and basketball. These are all good forms of exercise. But they may also want to learn some of these other sports which can also keep them healthy.

Exercise is important, but the body is only part of the person. We also have a mind which needs to be trained. And a soul which needs strength.

Our soul is the part of our personality which lives forever. It needs to listen to Jesus Christ. It does this best if we practice by doing it often.

"Bodily exercise is all right, but spiritual exercise is much more important and is tonic for all you do. So exercise yourself spiritually and practice being a better Christian, because that will help you not only in this life, but in the next life too" (1 Tim. 4:8).

1. Why is gymnastics so healthy?
2. What is your favorite sport? Why?
3. How can we exercise our soul?

My body is important and I can't let it become rundown. My soul is more important and I need to let God speak to it.

seven

How Babies Begin

Most of us know where babies begin their life. They start in the mother's womb. After spending around nine months inside her, the baby is ready for the outside world.

But what happens in the mother to make the child begin? It is an interesting story of God and nature creating life.

Two tiny parts are needed to make a baby. They are too small to be seen well without a microscope. One part is called a sperm cell. It comes from the father. The second part is the egg cell. It is made by the mother's body. Children are made by both the father and mother.

If we could see a sperm cell, it would look like a tiny tadpole. It has a small tail and swims quickly. The sperm wants to find an egg and begin a baby. However, of the billions of sperm not many match up with an egg.

When these two cells come together, they fertilize. This means the tiny parts have started to make a life. The two cells grow by splitting and still staying together. The two become four, the four become eight, eight become sixteen.

They keep splitting and growing for eight days and yet it is only a small speck. You could put fifty of these on one inch of a ruler. When it is this size, the cells rest against the mother's womb. There it will stay, grow and live until it is time to be born.

The sperm and the egg are good for only 48 hours. If they do not meet and begin a baby within two days, they die and new ones are needed.

The minute these two cells meet most of the child's looks are put together. Its hair color, sex, face and more are all settled.

As the cells grow they form three layers. The outside layer forms the skin. The middle layer makes the skeleton and nervous system. The third layer will create our digestive tract.

Scientists have studied the making of a body and know a lot about it. But still many things are a mystery. But all of life is filled with the unknown. How does the wind begin? Why do the stars stay in place? How can birds find their homes after winter has left? Why do salmon return to a certain pond?

There are also many things we do not understand about God. Where does he live? What does he look like? Where is heaven?

Without understanding all of this, we still have faith in God. Anyone who could create a way for children to be born must be a great God. He is worth following.

"God's ways are as mysterious as the pathway of the wind, and as the manner in which a human spirit is infused into the little body of a baby while it is yet in its mother's womb" (Eccles. 11:5).

1. What two cells make a baby?
2. How long are these cells able to make a baby?
3. Why do you believe in God?

It is easier to believe in God when we look at our bodies because the whole idea was His.

eight

The Body's Biggest Hinge

Children often enjoy making their knees jump. They will sit in a chair and let their bent leg hang loosely. Then a friend, parent or doctor will tap it lightly. If it is done correctly, the child's leg will jump. This happens because we have nerves which jerk when tapped. The real name is the patellar reflex.

The knee is particularly important for boys and girls because it allows them to run, jump and kick. Try running with your knee stiff and see how much you miss it.

Our upper and lower leg bones meet at the knee and look like a door hinge. They fold back the same way a door swings. Fortunately for us our knee bends backward but not forward. If it bent both ways we would spend all day trying to balance ourselves.

To stop it from bending forward God placed a kneecap at the hinge. It measures the size and shape of a chestnut. When you stand up your upper leg straightens against the kneecap. It holds the leg in place until you decide to bend it again.

We often hear of football, baseball or basketball players who injure their knees. Our knees are held in place by muscles and ligaments. Each knee has four main ligaments which work as tiny ropes to hold everything together.

If a quarterback has his knee twisted badly, he may injure his ligaments. When they are torn too much the athlete will need surgery immediately. Some players have to give up sports because their knees are hurt too seriously.

When an athlete has injured his ligaments, he has to work hard to strengthen them again. This usually includes hours of tough exercise.

In between the two leg bones is a cushion called cartilage. It holds body liquid like a sponge. Cartilage stops bones from rubbing against each other. Without this careful "oil" system our body would wear out in a few short years.

There is one comical scene in the Bible about a man's knees. After Belshazzar saw a man's finger write on his wall, "his face blanched with fear, and such terror gripped him that his knees knocked together and his legs gave way beneath him" (Dan. 5:6).

In the future all of us will put our knees to extra good use. One day every human being will bend his knees before Jesus Christ. We will bow down and call him Lord.

"That at the name of Jesus every knee shall bow in heaven and on earth and under the earth" (Phil. 2:10).

1. What does a kneecap do?
2. What is a patellar reflex?
3. What will we all do some day?

Someday we will be happy to bend our knees to praise Jesus Christ.

nine

Covered with Muscles

If we ask a small boy to show us his muscles, he usually makes a fist and draws up his arm. Most children think of a muscle as the bulge on our arm. But actually our entire body is covered with over 500 muscles. Not just a lump here and there. Muscles spread across us like a sheet on a mattress.

Our arms, legs, hands, feet and face all have a thin layer over them. Muscles come in different sizes and shapes, but each has an important job. Our bones would not be able to work if muscles did not control them.

It isn't easy work to be a muscle. They don't get a vacation or even a nap. While we sleep at night they keep doing their duty. If they quit, our face would droop and we would stop breathing.

To feel your muscles at work merely open your mouth. Your lower jaw dropped, but not by itself. It didn't just fall but was carefully lowered by muscles. When you close your mouth, stronger muscles lift your jaw. These are not the same muscles which simply go up and down. One set lifts and the other lowers.

When we go outside it will take 300 muscles working for each step. If these muscles didn't help each other, we would lose our balance and land on the ground.

The muscles surrounding our ribs work all the time without getting tired. If they got sleepy, we would have to stop breathing so they could rest.

Other muscles in our arms, legs and neck can be used only so often. After repeated use they get tired and have to quit for a while. You can feel it after playing baseball or taking a hike. Yet our muscles are tough and soon regain their strength.

If we practice using these muscles, they can be used for longer periods of time. Sometimes they even grow larger.

Muscles are called voluntary and involuntary. The voluntary ones work when we ask them to. Our arm muscles are voluntary. Our involuntary muscles include the ones in our face. They work most of the time without being asked.

People enjoy health and good muscles. It's fun to be strong enough to play, run and work. Not everyone has a strong body. But there is one thing all of us can have, and it's more important than muscles. Each one can have a good respect and love for God.

To follow God will make a person stronger than the largest muscles in the world.

"The speed of a horse is nothing to him [God]. How puny in his sight is the strength of a man. But his joy is in those who reverence him, those who expect him to be loving and kind" (Ps. 147:10, 11).

1. What is a voluntary muscle?
2. Which muscles work while we sleep?
3. How does God give us strength?

There are many things we can't do. But we ask God to help us with His strength.

ten

The Fire Inside

If our body is to grow and stay healthy, it has to have certain foods. The taste buds in our mouths enjoy ice cream, chicken, and cake. But our body needs what is inside these delicious foods. It would like the protein, fat, sugar and starch which these foods carry.

When someone eats a piece of cake with thick chocolate frosting, his body is glad to get it. It picks out the sugar and starch which it needs. Our body uses them by burning our food.

This doesn't mean there is a flame like a campfire. But there is heat which burns up our sugars and starches. That is why our body has a temperature and is always giving off heat.

It works a little like a campfire. A fire has to have fuel or something to burn. In this case it is parts of food. A fire must also have oxygen which comes in the air. If a stove has no oxygen, it won't burn.

When we breathe, our body takes in oxygen and our fire keeps going. We run out of fuel (food) so we have to stop and eat another supply.

Both the oxygen and the foods make their way into our blood system. They travel throughout our body and make the energy we need.

It is important to keep our food intake balanced. If we don't eat enough fruits or vegetables or eat too much cake, excellent bodies may begin to do strange things.

There isn't much scurvy in North America now, but there used to be. People who did not eat enough fruits and vegetables became sick. They would get large red spots on their skin or their gums might become swollen and sometimes bleed. Today parents and schools advocate these foods and our health has greatly improved.

Another disease which used to be common is rickets. When children did not get enough vitamin D they would often have soft bones. If a child has rickets his bones can be bent rather than straight.

Today this disease is rare because vitamin D has been added to milk. If we eat well-balanced meals, most of us get good supplies of vitamins.

Eating well is one of the reasons people in North America live longer than they used to. If you were born in 1910, you were expected to live to be 48. If you were born today, you are expected to reach 80.

The wise person is careful what he eats. We all like some "fun" foods, but too much can be harmful. To enjoy life at its best we need to keep our body as healthy as we can.

"The good man eats to live, while the evil man lives to eat" (Prov. 13:25).

1. Describe the fire inside us.
2. What causes rickets?
3. Explain Proverbs 13:25 in your own words.

The body God gave me runs well if it has the right fuel. Remind me to not stuff it with too much "fun food."

eleven

Do We Need Two Eyes?

Most animals get more use from their eyes than we do. They don't have to look at a tree with both eyes. Their left eye can watch a tree while the right one looks at a dangerous bobcat.

But when it comes to people, our two eyes have another purpose. Each eye sees just about the same thing. With two eyes we are better able to see how far away an object is. We can find its exact location.

Try an experiment. Hold a stick at arm's length. Now close your left eye and see where the stick is. Next close your right eye and look with your left. Did you see the stick change positions? Now look at it with both eyes. You can see the stick where it really is—in the middle. A human being can see better with both eyes.

Since our eyes do not look in two different directions at once, God gave us a neck. It quickly moves from left to right so we can see anything to our side. An owl can move its neck to see things behind it.

The human eye often does an excellent job of picking out colors. Most animals are not nearly as good at that. Many times we get a laugh from saying someone is color-blind. This means there are a number of colors which they can't make out. Often they have trouble with traffic lights because they can't see the difference between green and red. However, if they know the top one is green, they wait for it to get bright.

Color-blindness is common among men but rare with women. Out of every twelve men one of them has trouble making out colors. Once in a while there is a person who is completely color-blind. To them everything is black, white or some shade of gray.

Each of us has a certain color in our eye. We probably have blue, brown or green eyes. The colored part is called the iris. Iris means rainbow. If our eyes are blue or green, bright lights probably make us squint more than if they are brown.

All of us who can see are thankful for our eyes. There are so many interesting things to see. In the future God promises us even greater things than we have ever seen. Maybe they will be in this life and possibly in eternal life.

"That is what is meant by the Scriptures which say that no mere man has ever seen, heard or even imagined what wonderful things God has ready for those who love the Lord" (1 Cor. 2:9).

1. What can an animal do with its eyes which we cannot?
2. What does color-blind mean?
3. What do you think you'll see in heaven?

Thanks for eyes to see the wonderful world you made. And thank you for the things we will someday see which we can't imagine now.

twelve

The Stories on Our Faces

A person's face is as busy as a forest at night. Our eyes dance from side to side. Smiles and frowns make our cheeks jump up and down. Some people seem to talk with their eyebrows. Others turn red easily.

If someone is a member of our family, we can probably read his face without much trouble. Maybe when brother is telling a joke he gets a twitch in his cheek. When dad is angry, his forehead may look firm. If mother has good news, her eyes are probably sparkling like a star.

Often children are more open and honest with the stories on their faces. As they grow older they may try hard to control their faces. When they feel sad, adults may smile for their friends. If adults think something is silly, they might try to look serious.

Sometimes parents can read a child's face when he is sleeping. A smile may mean a pleasant dream. A harsh frown could show a nightmare.

A face has only fifteen bones with limited movements. The real stories are told by the muscles under the skin. They are bouncy and quick to tell how the mind feels.

Higher levels of animals do the same thing but not as well. A dog can look sad or bright and happy. A cat can curve its lip to snarl when it is angry. Monkeys are able to make the same facial changes. Some people suggest small insects show feelings on their faces, but people can't see them.

Usually people are more fun to be around if they have happy faces. It doesn't have to be bright and cheery all of the time, but surely much of it. The face which is grumpy, twisted or frowning is one we usually avoid.

We also notice many facial expressions affect other people. Sometimes our sad look makes another person look sad. Likewise, it is hard to be around a smiling person long before we start to smile also.

When we see black and blue marks on an arm, we know it has been hurt under the skin. When the skin is clear, the arm is probably healthy.

A face is like a bruise. A happy face usually means a happy person. Someone always in trouble often looks worried or troubled.

"How wonderful to be wise, to understand things, to be able to analyze them and interpret them. Wisdom lights up a man's face, softening its hardness" (Eccles. 8:1).

1. How do your parents' faces look when they are upset?
2. Describe the bones in your face.
3. Give three reasons God has given us to smile. (Make up your own.)

Thank you for giving our family reasons to be happy. As we count the good things we have, cause our faces to smile.

thirteen

Why I Get Hungry

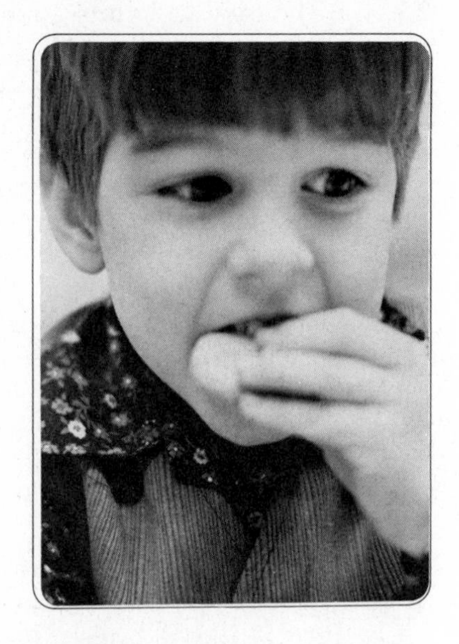

Which one sounds the best to you? Ice cream, potato chips or chocolate cookies? Even if you have just eaten a meal, usually you could still make room for one of these.

A person doesn't have to be hungry to enjoy some foods and drinks. Just thinking about them makes us want to eat. They set off a part of our body called an appetite.

When we see or smell our favorite food, water begins to move in our mouth. At the same time fluids, called gastric juices, start flowing in our body.

Babies begin their life with hunger, but probably not much of an appetite. As we grow, some foods taste better than others. We are training our minds to like some foods and reject the next kind. Therefore our sight or smell will send a quick message to our brain. The message says, "Man, root beer smells delicious." As quick as a wink our brain contacts the rest of our body.

This is when our juices start moving in the mouth and stomach. They are saying, "Send us some root beer." But still we aren't hungry. We don't need the drink or food; we merely want it.

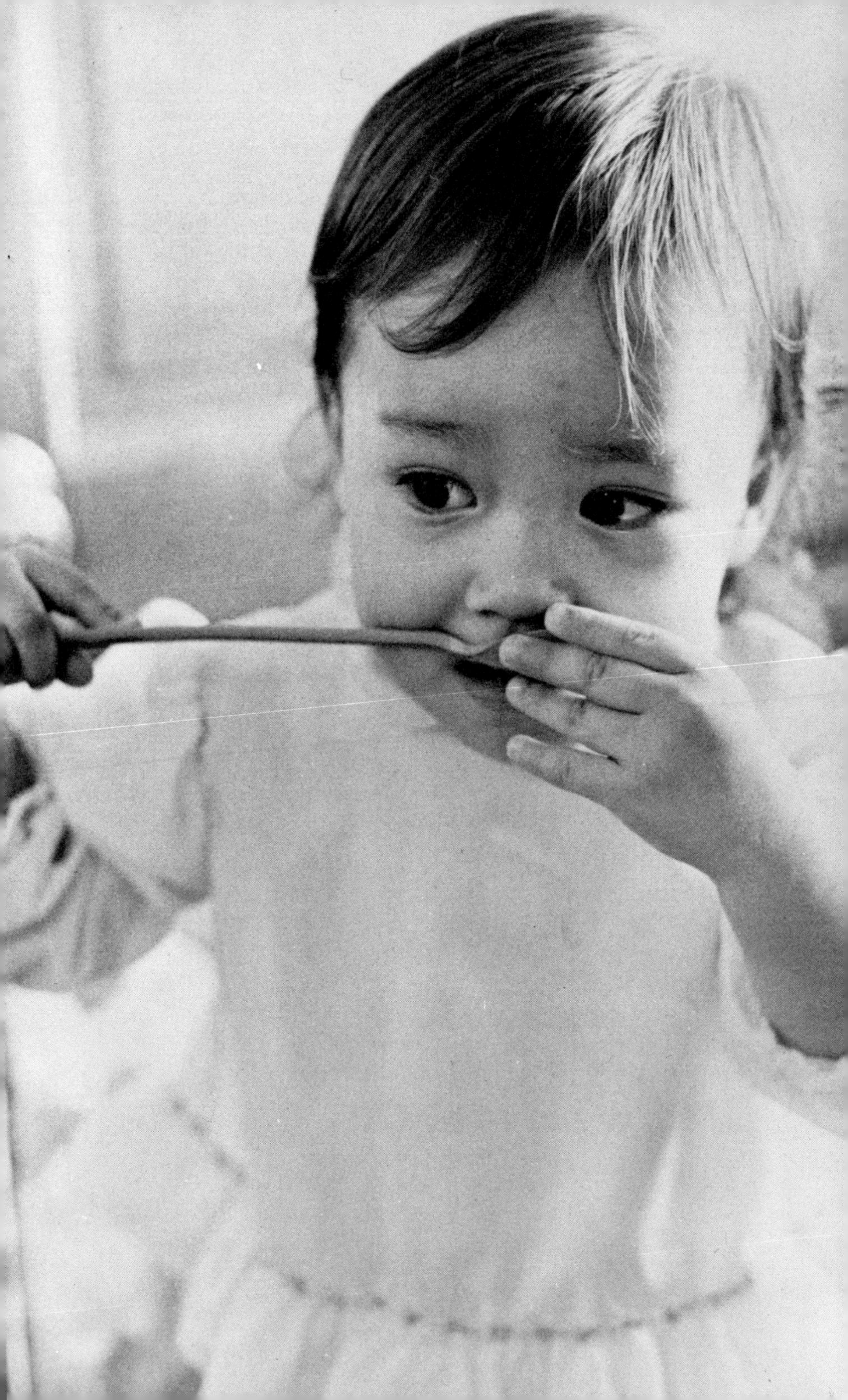

Ivan Pavlov, a famous Russian scientist, proved that our gastric juices will start flowing at the sound of a bell. He fed a dog each time a bell rang. The dog got used to it and his juices would begin each time the bell rang—even if there was no food.

Hunger is different because it means our bodies need food. Normally we think of hunger coming from an empty stomach. Sometimes this is the case. However, we can have a full stomach and still look for more food.

If our bodies don't eat enough of the right food, we can be "starving" even though we are stuffed. The body will send messages to the brain saying, "I don't have enough vitamin B down here," or "I need more iron; how about some liver?"

In turn our brain gets the word to our gastric juice: "Let's get some food."

These are systems which work every day in our bodies. We don't have to ask them to work or worry about them. They let us know if our food supply is low. They even get us ready for some good ice cream which we might not need.

Jesus Christ knew we needed to know more about God. We were "hungry" for Him. Jesus came to earth so we would be filled. We can meet God by meeting Jesus Christ.

"Jesus replied, 'I am the Bread of Life. No one coming to me will ever be hungry again. Those believing in me will never thirst' " (John 6:35).

1. What is the difference between hunger and appetite?
2. How can you eat and be starved?
3. Who can meet our hunger for God? How can He?

We need to meet God. Thanks for sending Jesus so we could meet Him.

fourteen

Burned Fingers

When we touch a hot pan our fingers bounce off quickly. The heat pushed against the tiny nerve endings and caused pain. The pain sent a special delivery message to the brain and our fingers snapped back.

The human body is covered with almost four million nerve endings to protect us. Of these, 200,000 are to check out the temperature. Half a million for touch and pressure. Three million more cover us just to send messages about pain.

Try to think of our body without them. A hand could burn in a fire because we didn't feel anything. We could freeze because we didn't know it was cold.

Cats are able to do the same thing through their whiskers. These stiff hairs vibrate when they touch an object. They "tell" the animal if danger is close enough to hurt him.

Our nerve endings are able to get used to some pain. If they didn't, life would be hard. For instance, when we get into a hot bathtub the heat hurts. But if we stay in it for a while, our body gets used to it. The nerve endings change in those few moments until the water doesn't bother us.

If this didn't change, we wouldn't be able to swim in cool water. It also explains why our clothing feels comfortable so we don't even know it is on us.

The touch and temperature nerve endings change more easily than the ones for pain. We would like to get used to pain so it wouldn't hurt. However, the pain needs to stay there for our protection. Usually a toothache doesn't go away until we get medicine or other help. God gave us pain to warn us of danger so we would do something about it.

Sometimes pain likes to play games. When we have a headache our eyes may hurt instead. If our appendix is in trouble, the pain may be felt near the breastbone. Some injuries give little pain and therefore are all the more dangerous. Our brain can be hurt and give no pain. The injury is often seen in the eyes or heard in the speech.

Not all pain is physical. Sometimes we hurt in our heart because we are disappointed. Maybe our dog died or our family vacation was called off. At other times we have done something bad and we feel sad about it. One day the psalm writer felt terrible because he did something he should not have.

"See my sorrows; feel my pain; forgive my sins" (Ps. 25:18).

1. How do nerve endings help us?
2. How do some pains fool us?
3. How can you share your pain with God?

When we do something to hurt someone else, whether we hurt our friends, brothers, sisters, parents, ourselves or God— thanks for forgiving us.

fifteen

Why People Faint

A boy was standing on a church platform one hot July afternoon. The children were practicing for their Friday night program. Suddenly his eyes started to close and then open, and his body began to sway. Quickly teachers came running to his side just as he fell to the floor like a dirty shirt.

People of all ages can faint. The whole process is simple. Our neck has two arteries (like pipes) which carry blood to our brain. Every minute a pint and a half has to be pumped there. Most of it goes through these pipes and some travels along our neckbone.

If something stops the blood from getting to the brain, we will pass out or faint. Being choked can do the same thing.

In the case of the boy on the platform, several things may have happened. He probably had his body too stiff cutting off his blood movement. He possibly was weak before lunch and the heat of the day took more strength away.

A person can faint because he gets too excited. Suppose a girl is frightened by a mouse. Her heart may speed up because she is scared. However, if she doesn't run or move fast, her blood remains slow. The result is a body calling for faster blood and not getting it. Therefore her mind isn't getting enough and she will pass out.

The pipes (arteries) leading to a woman's brain are smaller than a man's. So it may be true that women faint more easily than men.

For older people the bathtub can be doubly dangerous. The hot water will make arteries slightly larger. These "pipes" call for more blood, faster. Since the body is resting, it doesn't send the blood rapidly. When grandfather starts to stand up his body is weak and his head slightly dizzy. He may faint and fall to the floor.

When children find out about these two arteries, they sometimes try to make another person faint. This is dangerous and someone could be seriously hurt.

Christ never would have done this. He wanted to help people and even worried about them fainting. One afternoon after speaking to a large crowd, he refused to let them go home until they had eaten. Jesus was afraid they would faint as they walked.

"And if I send them home without feeding them, they will faint along the road! For some of them have come a long distance" (Mark 8:3).

1. Describe what can happen in our body to make us faint.
2. Why do some girls seem to faint more easily than boys?
3. Name some little things in our life which Jesus Christ cares about.

Thank you for showing your love for people. This is how I know you love me.

sixteen

One-Pound Pump

If someone wanted to send water through a pipeline almost 100,000 miles long, he would probably need a large powerful pump. The human body sends blood through our system, which is just that long. To do it our body uses a small heart about the size of our closed fist.

The simplest way to describe a heart is by its use. It is a human machine which has two pumps. One side pumps a bluish red blood which has come from our food and oxygen. This blood is pumped into our lungs where it gets rid of carbon dioxide and gains more oxygen. Then the second pump takes over and pushes the new blood into our system.

Fortunately for us, this little lightweight machine is steady and tough. The heart has about ten pints of blood to work with, and day and night it pumps basically this same blood around and around our body. During one day our blood will make over 600 complete trips through the body.

Normally the heart beats 70 times each minute or slightly faster than once every second. To a large extent we each control our heart. For instance, if we run fast our heart may increase to eight times its usual speed.

If we have a nightmare our mind sends messages to our heart. We then wake up with it pounding very rapidly even though we have been resting.

While good exercise is valuable, too much at the wrong age can also be dangerous. An athlete's heart may swell much larger in size because he exercises. However, the person who is not used to running is asking for trouble if suddenly he runs too much. Obviously, if our heart stops beating we will die. If it stops for only a few minutes and then starts again, the lack of blood flow to our head may cause brain damage.

Many things could be dangerous to the human heart. People who are tense most of the time give the heart a heavy burden to carry. Smoking also makes parts of the body sluggish and may make the heart beat 80 times per minute rather than the regular 70. The added load may prevent the heart from lasting years longer. Eating the wrong foods can do the same thing.

For many years people believed the heart was the center of our emotions. We say, "He broke mother's heart"; or a sad person's "heart aches." It is true, strong emotions may affect our heart. But usually it is a figure of speech and everyone knows what we mean.

Jesus used the heart in a figure of speech by referring to it as the center of a human being. He taught us if we tied up a bad boy, he would still be bad. But if he changed inside, from the heart, he would start to become a good boy. Jesus changes us from the inside. He changes what we want, how we feel about people, what makes us angry. He changes our "heart" and not just the outside.

"But evil words come from an evil heart, and defile the man who says them" (Matt. 15:18).

1. How much blood is in most bodies?
2. How fast do most hearts beat?
3. What did Jesus mean by an "evil heart"?

If we feel right about people, we will treat people right.

seventeen

The Busy Tongue

Animals use their tongues in many ways. Cats use them to lap up water. Anteater tongues pick up food, and dogs us theirs to cool themselves off.

People can use their tongues for a variety of purposes, but it has one main job. It allows us to make the sounds necessary in speaking. When kings didn't want someone to talk anymore, they sometimes had the person's tongue cut out.

To feel how important the tongue is, try saying a few words. Say "the." We can feel our tongue press against our teeth. Now try "get." Our tongue has to push up to make the "g." Then it has to hit the roof of our mouth to smack the "t."

Without thinking about it, the tongue is always dancing around in the mouth. When we whistle it rolls into position. If food gets caught in our teeth, it tries to become a toothpick. When a tiny piece of paper gets in our mouth, our tongue becomes an instant policeman to arrest and throw the stranger out.

Sometimes our tongue takes the place of words. If we don't like something, we might stick out our tongues, even though it isn't nice. When we want Mom to know the ice cream is especially good, we might merely lick our lips.

Most tongues are only four inches long (though some are double this). They may weigh a slight two ounces. But they are quicker than darts.

When it isn't busy talking or clearing out our mouth, the tongue has other important jobs. It plays an important part in helping us taste food. There are taste buds which cover our tongue. Our mouth is filled with these buds and our tongue is a valuable helper.

They are called buds because they look like the tiniest rose buds. We can't see them without a microscope. Inside the buds are even smaller parts called "cells." These send messages to our brain to tell us if the food is good or bad.

It is true that all of us do not taste things the same way. Candy might be too sweet for one child and just right for the next. Salt may be too strong for one and delicious to the other. However, our buds can also be taught to like some tastes.

Another helpful purpose of the tongue is to tell us when we are sick. If it is too red we might have one type of illness. When it is yellow we could have another sickness. However, most of us can't really tell from looking at our own tongues. We might be just looking at the coloring left by purple jelly beans. Only a doctor knows how to "read" our tongue for certain.

The tongue can be a great help in many ways. It can also be a cutting sword. When we use it wrongly to stab people with words, it becomes a painful weapon. Today is a good time to say nice things about our friends. It will make their life happier.

"Anyone who says he is a Christian but doesn't control his sharp tongue is just fooling himself, and his religion isn't worth much" (James 1:26).

1. How many uses can you name for a tongue?
2. What is the job of the "rose buds" on our tongue?
3. Why do Christians have trouble controlling their tongues?

Today I could say mean things about people or I could say some kind and loving words. If I say the kind ones, I can help my parents, brothers and sisters have a better day. Lord, help me to be kind.

eighteen

Where Do Children Come From?

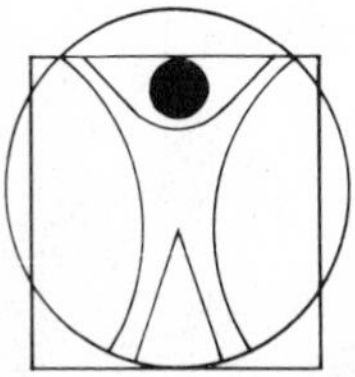

When a child is born he is usually already nine months old. It was when the sperm and the egg came together inside the mother that a child started to form.

God gave mothers an excellent place where life is created and turned into a child. It is called a uterus or womb. It is a pear-shaped area and it takes care of the child until it is ready to enter the outside world.

Soon after the sperm meets the egg the egg sends out a tiny feeler and attaches itself to the wall of the womb. Over the next eight or nine months this line will supply everything the unborn child needs.

The line is called an umbilical (um-bil-i-call) cord which may be anywhere from five inches long to four feet. As the egg grows into a person it gets all of its food through this cord. It also sends off its waste material the same way. It uses its mother's kidneys and lungs to keep alive. The oxygen, vitamins and mineral supply all come from the mother also.

All of us have a "belly button" or navel. This is where our cord was connected before we were born.

As the child grows larger the womb will grow with it. The muscles grow stronger with it in order to hold the extra weight.

The unborn child moves about in the womb to some degree until around the seventh month. Afterwards it may move arms and legs but it takes a fixed position and waits to be born. Birth usually comes after nine months. But many children are born much earlier or later.

After birth the womb returns to its regular size. It is then prepared to possibly house another unborn child.

Around the age of 40-42 many women go through a change called menopause. Then their womb returns to the size a little girl has and they have no more children.

Like the rest of our body God has made the womb as a carefully created home for a tiny delicate unborn child. There it is protected and provided with everything it needs. As the psalm writer looks at his life he is thankful God has protected at all stages in his life. He realizes God became the watchman of his life even before he was born. Now he asks God to keep up his protection and guidance even while he is older because he knows he still needs it.

"Lord, how you have helped me before! You took me safely from my mother's womb and brought me through the years of infancy. I have depended upon you since birth; you have always been my God" (Ps 22:9, 10).

1. What is the purpose of the navel cord?
2. Where does the child live before it is born?
3. What did God do while you were waiting to be born?

Thanks for watching over me from day one.

nineteen

How We Smell

If our nose were completely dry, we wouldn't be able to smell anything. In order for a human to smell, millions of tiny particles have to leave the flower or food and travel into our noses. Inside they make their way to the upper part which is near our eyes.

Once there they land on a small (postage stamp size) pad covered with very little hairs. This fuzzy area is upside down and moist. When the particles from the rose land here they become wet and dissolve. Only now can we smell them and either enjoy or dislike it. This is called our olfactory system.

Animals have a better sense of smell than we do and they use it more. They sniff to identify their home, their food and even friends. A male gypsy moth can smell another gypsy moth miles away.

It would appear the ability to smell is a very simple thing in humans, but even this is amazing. The small pad we spoke of contains ten million receiving cells. Each tiny cell is protected by six to eight hairs. As soon as the pad receives an odor it sends a quick message to part of the brain. This message goes to the brain and returns at the speed of 350 feet per second. That makes it 21,000 feet per minute.

Smell remains important to us both for pleasure and also to warn of danger. A large part of the enjoyment of a pizza is being able to smell it. Natural gas doesn't have a smell and therefore could be dangerous. It might leak and no one realize until it is too late. Odor has been added to gas so we can smell it and sense trouble.

Man is not as dependent on his sense of smell as animals are. Therefore many creatures have bigger smell lobes in their brains than humans have. We have not begun to use the full power of smell. Many people can identify approximately 4,000 scents. However, man is capable of learning 10,000. As we grow older our sense of smell gets duller as do many of our other senses.

There are a lot of things we don't know about God, but the Bible tells us he has many of the same senses we have. He can touch, hear, see and yes, even smell. And why? Because he is alive.

Some people worship idols such as trees or rocks or even the sun or moon. Yet, they can't be real gods. They aren't alive. They can't react and the psalm writer says they can't even smell.

If we are going to worship God, then by all means let us worship the one who is alive.

"For he is in the heavens and does as he wishes. Their gods are merely man-made things of silver and gold. They can't talk or see, despite their eyes and mouths! Nor can they hear, nor smell, nor use their hands or feet! Nor speak! and those who make them and worship them are just as foolish as their idols are" (Ps. 115:3-8).

1. Describe a smelling pad.
2. Why can't we smell more scents?
3. Can God smell oven-baked bread?

Thanks for sending Jesus Christ to show us God is definitely alive.

twenty

Tough Teeth

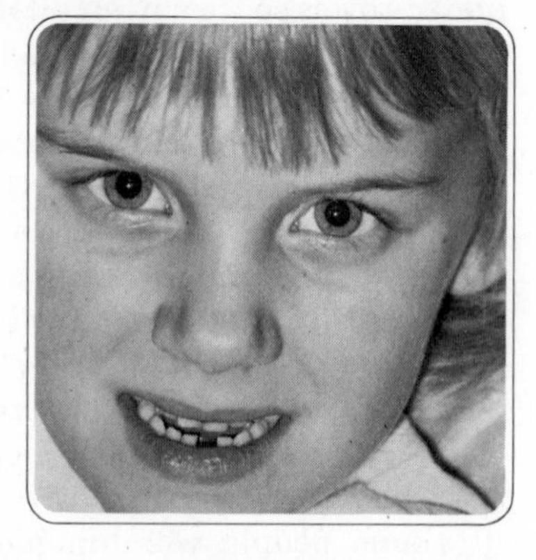

When we become adults most of us will have 32 permanent teeth. However, for one reason or another a lot of people don't. Four teeth, called wisdom teeth, usually wait until we are in our twenties before they decide to join the bunch. About one person out of every ten is born without a few adult buds, so he won't get 32. Then, of course, many people let their teeth decay with cavities and they don't end up with all they should.

Actually teeth are very well made and normally should last us all of our lives. Teeth are the hardest materials in our bodies. The enamel which covers them is very strong; however, it doesn't grow back. Bones are excellent at repairing themselves with a little guidance from the doctor. But when teeth are decayed, chipped or removed they need to be filled or replaced.

The biggest enemy to the enamel is the tiny bacteria which is often left from food. The bacteria eats away at the enamel unless we first remove it by brushing regularly.

Underneath the enamel is a large section called dentin. Related to our bones, this area can be very sensitive. If decay is allowed to get through the enamel into this second section, pain may be the result. Dentin is like ivory and makes up the bulk of the tooth.

Inside the dentin is the dental pulp which contains nerves and blood vessels. If tooth decay is allowed to go as deeply as the pulp area the person may be in for difficult and painful times.

Tooth decay is probably the most common health problem man has. Practically everyone has trouble with it sometime in his life. Half of all the three-year-old children in the United States have at least one cavity. Fortunately, at 35 the decay system starts to slow down, but then some people have trouble with their gums.

Two of our side front teeth got their name in a strange way. They are called eyeteeth. People used to believe these teeth had such large roots that they went all the way up into our eyes. They were afraid to get them pulled because they thought they would go blind.

Dentists tell us one thing modern man fails to do is to exercise his teeth. In a day of easy-to-eat soft foods he needs to practice chewing foods such as apples, celery and the like.

Anyone who has ever had a toothache will find Proverbs 25:19 easy to understand. When we are undependable and keep breaking our promises to our friends, we are a big pain—just like a toothache. We don't like to have people let us down and we need to try to be as good a friend as we possibly can. God is easy to appreciate partly because he is dependable and keeps his promises.

"Putting confidence in an unreliable man is like chewing with a sore tooth, or trying to run on a broken foot" (Prov. 25:19).

1. What is the biggest enemy to our teeth?
2. How did eyeteeth get their name?
3. Are you known as someone who keeps his promises? Explain.

We need to be known for our dependability and not for our broken promises.

twenty-one

There Is No One Like Me

Every person is different. Of all the billions and billions of people who have lived, no two have been the same. Even if they are identical twins, it is possible to tell them apart.

No two people can write exactly alike. If we try to copy someone else, our handwriting may look like our brothers, but it is different.

The same thing is true of fingerprints. The tiny lines on the end of our fingers are different from anyone elses. The F.B.I. has over 100 million sets of fingerprints and no two look alike. The Chinese used to sign their names by using their fingerprints. They simply put their thumb or finger in ink and pressed it on the letter or paper. They knew only one person had that print.

Often when a baby is born a print is made of its foot. The lines on its sole are like no one elses. The same thing can be done with the palm of the hand.

Our speech has the same special mark to it. Even though we are born of the same parents, live in the same home and eat the same food, our voice will sound different. So different that no one, even the best pretender, can say a word the way we do.

A famous scientist claims you cannot say a word twice the same way. Maybe we can't hear the difference, but it is there. If you said "sport" fifteen times or said it all day, it would change slightly every time. According to him, people have said "run" for hundreds of years and it has never been said the same way twice.

Not only do we look and sound different, no one else even smells like us. Sometimes when a person is lost, bloodhound dogs are used to help find them. They first sniff the lost person's clothing and then look for someone who smells the same way. They will not track down the wrong person because no one else smells like that.

If the lost person is dead, the hounds will have trouble finding him. It seems that our body stops giving off the scent when we die.

All of this makes us someone special. In a crowd or a small group no one is like me. I can never be just a number or another "kid." I fill up a place in this world which no one else will ever fill.

"If a man has a hundred sheep, and one wanders away and is lost, what will he do? Won't he leave the ninety-nine others and go out into the hills to search for the lost one?" (Matt. 18:12).

1. Describe three ways we are different from anyone else.
2. How can bloodhounds find people?
3. What does the story of the lost sheep tell us about Christ's love?

Thank you, God, for not being too busy to care about me.

twenty-two

Hearing Without Hearing

Not only are good ears necessary for hearing, but they also stop us from falling on our faces. Inside our head there is a mechanism called the "inner ear" which is next to our regular ear. It consists of three little loops which tell the brain if we are balanced while walking or are about to fall down. As quickly as the brain gets the message it tells the muscles to straighten up before the body gets hurt. People who have trouble with the inner ear have a very difficult life until it can be corrected.

The part of the ear which we all can easily see is important but not the most useful section. The outer ear really works as a sort of trap. As sound travels through the air this outward skin serves as a funnel to guide it into the middle or real ear. To see how helpful it is, merely cup a hand behind one ear and notice how the added flap makes listening easier.

Many animals use this trick and pick up their ears when they want to hear better. Years ago people who had difficulty hearing sometimes carried a small horn. By placing the small end to their ear they improved their hearing considerably.

Inside our head on either side is a small drum, just like the canvas on a drum in a band. When sound vibrates on this drum it pushes against a small bone called a hammer. The hammer pushes against another bone called an anvil and in turn pushes against a third called a stirrup. The last one is called a stirrup because it looks almost exactly like the side stirrup on a horse saddle.

From the stirrup the sound is carried to the tiny hairs on the cochlea. These hairs send the message of the sound directly to the brain, and the brain decides what to do about it. To help the brain get the message these three little bones increase the pressure of the sound by 15-20 times.

Briefly stated the three parts of the ear are:

(1) Outer ear for protection and trapping sound

(2) Middle ear to hear

(3) Inner ear for balance

As people get older some think their hearing isn't as good as it used to be, and they are probably right. After 50 years of age people often have trouble hearing high tones.

As with most of the body the ear has a good system to protect itself. Equipped with microscopic hair and wax these two stand firm to keep out small insects and particles of dust. The wax also makes a good seal when we decide to go swimming.

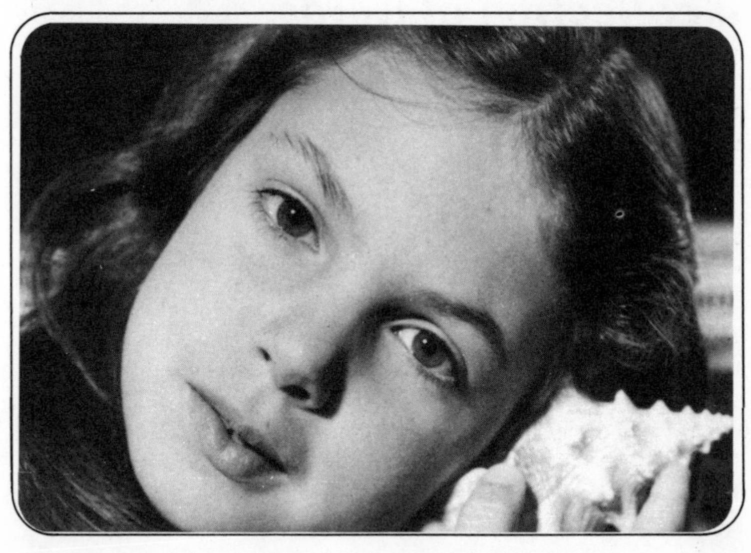

While some find this hard to believe, most experts agree that too much noise over a period of time can be very harmful. Therefore, radios and record players should be used carefully.

There are a lot of people who can hear but can't hear. Sometimes a parent will give instructions to their child and then say forcefully, "Do you hear me?" It's obvious the child heard. The parent is really asking, "Do you understand the importance of what I am saying?"

Jesus met a lot of adults who could hear without hearing. They had become stubborn and couldn't change to follow God. Some of us hear regularly from God in the Bible, but it doesn't help us because we hear without really listening.

"Blessed are your eyes, for they see; and your ears, for they hear" (Matt. 13:16).

1. Describe the three parts of the ear.
2. Explain how a hammer, anvil and stirrup work.
3. What did Jesus mean about hearing without hearing?

We need to stop and understand or else faith will be just a collection of words.

twenty-three

Windshield Wipers

Most windshield wipers we think of are the ones which move from left to right. The body has wipers but they move up and down and are called eyelids.

Very simply put, they work this way. Underneath the eyelids there are tiny tear glands which give off water. They keep the eye moist. The eyelids sweep up and down to wet the eye and at the same time push any particles out of the way. Therefore, the "windshield" of our eye is kept clean so we can see clearly.

Normally we don't have to ask our eyelids to move and clean our eyes. In fact, eyes blink several times each minute and usually without us noticing it. When we blink on purpose our lids move much slower than when left to themselves.

Eyelids also work well in preventive cleaning. Because they are so fast, they often block many particles or bugs before they can get into the eye. Those few specks which get past the gate are then quickly washed away.

While standing guard against intruders the eyelid plays a dual role in helping our vision. By moving our lids we can adjust the amount of light which enters our eyes. We are also able to alter our focus the same way.

Above our eyes is another line of protection called our eyebrows. The brows and the bone under them keep another selection of intruders from getting past. When we are sweating, an eyebrow often comes in handy to hold back the flood. When playing in a dusty basement eyebrows help stop falling particles from entering our eye.

The white part of our eyeball is covered with a substance called conjunctiva. It contains a large amount of nerves that serve as a warning system. If something gets near our eye the conjunctiva has a hair-trigger and sends off an immediate message to the eyelids to close at once. Some things get past this screen system, but very few.

Man probably wasn't around very long before he learned to use his eyelids for fun and games. If a boy wanted to tell a girl he liked her, he might just use his eyelids and wink at her. If a friend wanted to tell someone he was just kidding, he would wink at him.

At about the same time people learned to lie with their eyes. If one promised the banker he would pay the money back in one month, he might just wink at his friend sitting next to him and that meant he was lying.

The Bible says people who break their promises and cheat others by "winking" are really liars. The person who loves God will not try to trick another person because he knows it is important to be honest.

"Let me describe for you a worthless and a wicked man; first, he is a constant liar; he signals his true intentions to his friends with eyes and feet and fingers" (Prov. 6: 12, 13).

1. Name two purposes for eyelids.
2. What is the job of conjunctiva?
3. What is wrong with lying?

An honest person never has to worry about getting caught.

twenty-four

Making Milk
(Baby Food)

While robins need to go and search for food to feed their babies, God gave man a different plan. Mothers have a marvelous factory which makes food inside their bodies.

This process takes place in a woman's breast. They know exactly when to produce food and do it only when they have a newborn child to feed.

Normally the breast is merely a large type of sweat gland referred to as a mammary gland. Girls start to get them as early as eight years old and as late as 18. The majority of girls get them around 12.

Made up mostly of fat the breast is faced with the complicated job of turning blood into milk. To do this each breast has thousands of tiny glands which make milk. Each one produces the smallest droplet which moves toward special "ducts" something like a river. The breast usually has around 17 of these rivers. The 17 streams flow to the nipple at the end of the breast. There they meet and produce enough milk to supply for the baby.

If that was all the breast did for a baby, it would be a lot. However, it is so remarkably made the gland holds back the milk until it has given the infant another important liquid. The substance is yellow and has little nourishment. For the first four days the breast feeds this (called colostrum) to the child and it helps to clean out the baby's digestive system. It cleans out stuffy mucus and at the same time helps to fight harmful diseases such as measles, whooping cough and others.

During this time the baby will lose weight, but it is important to get this medicine. On day five each breast starts delivering half a pint of milk daily and as the child's needs grow, the flow also increases. In order to do this the mother has to contribute a great deal of her own blood.

While cow milk is good for children, mother's milk is even better for a baby. The exact vitamins and minerals are available in just the right amounts. Many mothers don't breast-feed their babies, but others are going back to the practice.

Most of us have watched a small baby go to sleep just after it has been fed. He usually looks very soft and comfortable. He has reason to be at rest and peace. He has just been cared for and personally fed from the breasts of his mother.

One day a man felt very safe and happy because he knew God was close to him. Looking for the best way to describe this great feeling, he said it was just like the peace a baby feels after drinking from a loving mother. He had learned to be still and trust God.

"I am quiet now before the Lord, just as a child who is weaned from the breast. Yes, my begging has been stilled" (Ps. 131:2).

1. How do "rivers" in a breast work?
2. What is the purpose of yellow colostrum?
3. What makes you afraid? Could God help you with it?

We get restless and afraid of tomorrow. Teach us to relax because God takes care of us.

twenty-five

Why Do We Laugh?

People do a lot of strange things. But few are harder to explain than why we laugh. There is no physical reason why we have to.

If we chose, we could show happiness by slapping ourselves on the back of the head. When we hear a funny joke, we could just pull our ear. Instead, people smile or throw open their mouth or make a noise or all three of these.

Small babies smile at an early age, and if tickled in the ribs people may giggle or even howl. Nevertheless, laughing at stories or actions is something a person must learn. Even the famous "laughing gas" doesn't really make people laugh. (Its official name is nitrous oxide.) When given to a patient it may make him feel light-headed and small pains are easier to bear.

Though it is difficult to explain, laughter is nevertheless real. When something seems funny to us our diaphram flutters up and down in our chest and we let out some sort of noise.

The facts seem to indicate that a person laughs for one basic reason: he wants to laugh. There is no joke that can make a person laugh. A person chooses to see something funny in it or he refuses. What will make one person laugh may even cause another person to cry.

A good example of this is a woman in the Old Testament named Sarah. When she was very old God sent a message that she would have a baby. Sarah considered it so strange that a woman her age could have a child she chose to laugh about it. Later, after her baby was born she laughed again. This time not because she thought it was strange but because she was very happy at what God had done. Abraham, the child's father, named his son Isaac because it meant laughter (Gen. 21:1-6).

Laughing seems to have a useful purpose in the human body. When people laugh they tend to relax and the strains of life seem less. This relief takes pressure off the body. Maybe this is the reason why the Bible tells us there is a time to laugh (Eccles. 3:4).

Laughing is a good healthy thing to do when done at the right time. God has given us the control over ourselves to decide whether we want to laugh or sulk and brood. God has certainly given us all plenty to be happy about.

"What happiness there is for you who are now hungry, for you are going to be satisfied! What happiness there is for you who weep, for the time will come when you shall laugh with joy" (Luke 6:21).

1. Why do we laugh?
2. How does laughing gas affect us?
3. How does God feel about laughing?

God gave us some things to laugh about because it's good for us.

twenty-six

A Puffed-Up Chest

Is there any truth to the story of women having more ribs than men? Well, yes and no. Most human beings have 24 ribs with 12 lined on each side of the chest. However, one person out of 20 has more than their two dozen. This happens to more women than men. Nevertheless, most people have the same number.

It's easy to explain how this could happen. We are born with around 300 bones. As we grow older many bones come together (fuse), and by the time we are adults there are 206. With some people not all of their bones go together, so they may have 210 or 215. Therefore we may end up with a few extra of this or that.

The ribs make up the human chest and serve a good purpose. They build a cage in which many of our important organs are stored and protected. Our heart, liver and lungs are all surrounded by this wall of ribs (called a thorax).

While they furnish protection from normal bumps or from easy puncture, the ribs aren't strong. They can be easily broken. But fortunately, as with other bones, they also mend well.

Each rib is connected to our backbone (vertebrae). In the front they are tied to our breastbone (sternum). The material which ties them in the front is called cartilage. It's easy to feel the breastbone and the adjoining ribs in the middle of our chest.

When we breathe in heavily our lungs fill with air. We can see and feel our chest get larger. When we breathe out our chest gets smaller. Normally we breathe 18 times each minute and our chest moves each time. The ribs are covered with muscles to give them added strength and protection.

Our ribs really come in handy if we want to look bigger or pretend to be extra strong. All we have to do is take a deep breath and swell up our chest. If we can hold our breath long enough, people might even think we are bigger than we really are.

Jesus talked about a man who was sorry for things he had done. He wanted God to forgive him. He didn't puff up his chest and act proud. He knew he was wrong and he was sorry. He very sadly beat on his chest and said, "God, be merciful to me, a sinner." He had stolen and cheated people and he wanted God to forgive him.

Sometimes we all puff out our chest and pretend we are tough. Adults do it, too. But there are other times when we should hang our heads and admit we were wrong.

"But the corrupt tax collector stood at a distance and dared not lift his eyes to heaven as he prayed, but beat upon his chest in sorrow, exclaiming, God, be merciful to me a sinner" (Luke 18:13).

1. As we grow, what happens to our bones?
2. Describe the major parts of our ribs.
3. Why did the tax collector beat on his chest?

Keep me from being too proud to say I am sorry.

twenty-seven

The Thumbless Kings

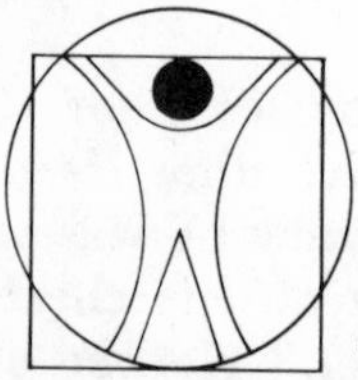

The human thumb doesn't sound like the most interesting part of the body. We usually don't think about it unless we hit it with a hammer or jam it while catching a ball.

To appreciate how important the thumb is, try an experiment. Take a piece of adhesive or masking tape and tape your thumb to your hand. Now see how many things you can do. Pick up a pencil, drink a glass of water. Try throwing a baseball or football.

It doesn't take long to miss a thumb. No finger would be missed as much because the thumb does 45% as much work as the rest of our fingers.

The thumb is very useful and yet not very complicated. There are 27 bones in the hand and the thumb only needs two. The four fingers need three each. The thumb rests in a saddle joint which allows it great freedom and flexibility. There are some animals with thumbs but without the good use human beings have.

If thumbs are important little parts, think for a moment about our big toe. Some animals use their big toes just like thumbs. Therefore, they can even eat with their feet. Not many human beings can do that, though sometimes a brother or sister can put his toe in his mouth.

The big toe is part of an excellent and many purposed system we call the foot. Each foot hits the floor hundreds of times a day and supports half our body weight. And yet the foot takes it all pretty well. The foot is called on to carry out an amazing trick. It has to balance a large body on a small area standing still, walking or running without tilting over. And yet the foot does the job and seldom complains.

The big toe, two simple bones like the thumb, plays more than its part. This toe carries one-fourth of our weight. If we weigh 80 pounds, each big toe carries and balances ten pounds. If we weigh 100, each carries 12 1/2 pounds.

The Bible tells us about a mean king named Adoni-bezek. He had fought many nations and had defeated seventy other kings. He ordered the big toes and thumbs be cut off each king he captured. Then they couldn't fight anymore. They couldn't hold a spear because they had no thumbs. They couldn't run or jump very fast because they had no big toes.

King Adoni-bezek wanted to make them look weak and have people laugh at them.

Then one day the Israeli army captured mean Adoni-bezek and cut off his big toes and thumbs. Unfortunately the cruel things he did to other people finally came back and happened to him.

" 'I have treated seventy kings in this same manner and have fed them the scraps under my table!' King Adoni-bezek said. 'Now God has paid me back' " (Judges 1:7)

1. How many bones are in the thumb?
2. Practice picking up an empty glass without using your thumbs.
3. What did King Adoni-bezek do to the seventy kings? What happened to Adoni-bezek?

We need to treat people as we would want to be treated.

twenty-eight

Bouncing Breath

What would life be like if breathing was work? Suppose we had to count our breathing: 1, 2, 3,—1000—2110. If we missed a breath, we might get weak or sick. We wouldn't be able to sleep at night. We'd have to stay awake so we could keep breathing.

It doesn't seem like a very happy job. Fortunately, God gave us a better system. Our body continues to take in air no matter what else we are doing.

Sometimes odd things interrupt our breathing. None is stranger than the uncomfortable hiccups. Our breathing becomes a jerky motion rather than its normal smoothness.

As far back as history goes, people have been getting the hiccups. Yet, we aren't sure what causes them or what makes them stop.

There is a thin sheet of muscles between our chest and stomach. It is called a diaphram. When we breathe these muscles move.

Something disturbs these muscles and causes them to jump or jerk. Maybe our stomach or intestines have become larger for a while and have changed shape. Our smooth ride has now become bumpy.

Nervousness, worry or shock can also cause our breathing to "bounce."

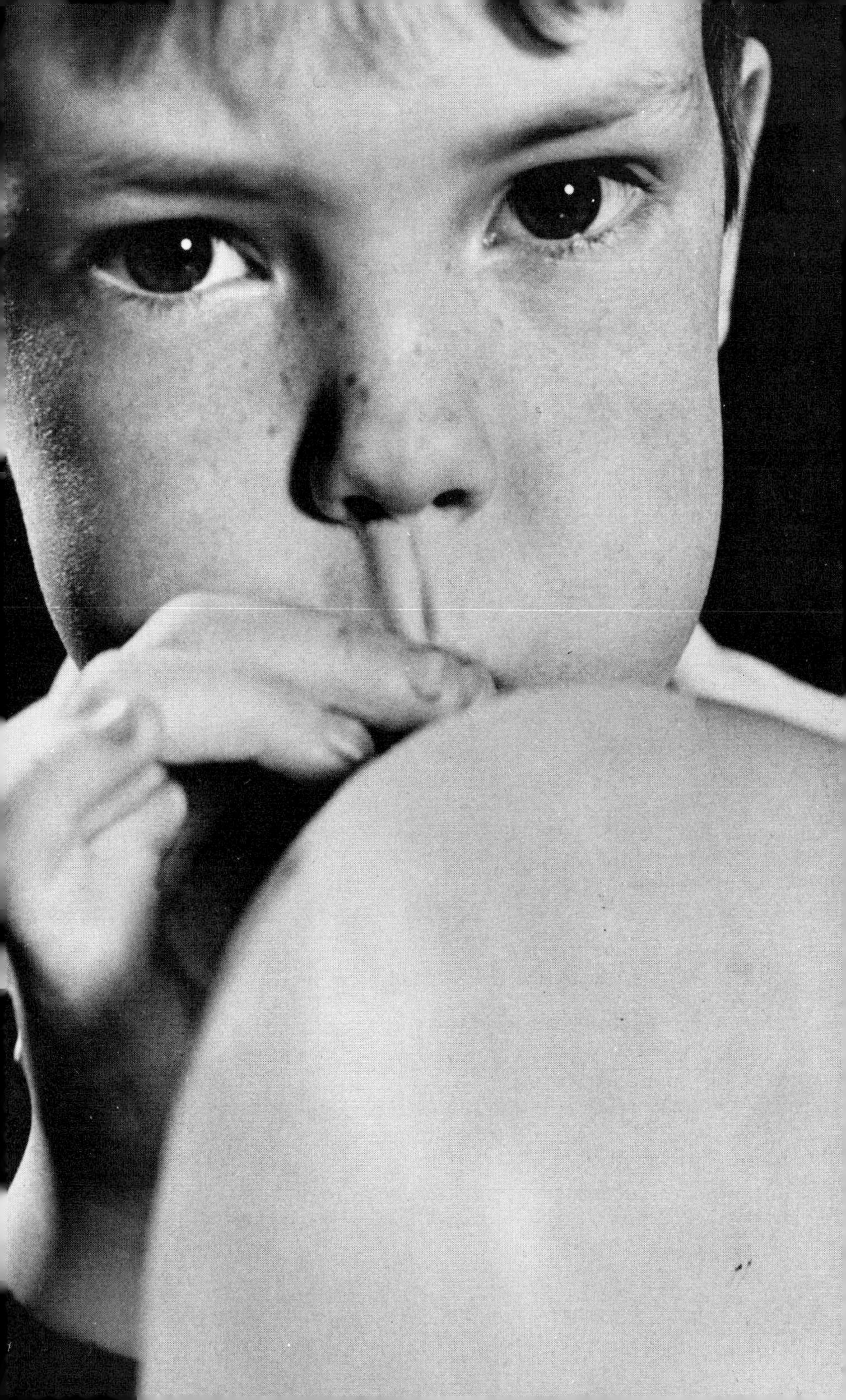

More mysterious than the causes of hiccups are the cures. Will holding your breath and counting to ten stop them? Will gargling with water bring them to an end? What about making yourself sneeze? Maybe any or all of these remedies will work. They are the same ones Plato suggested four hundred years before Christ was born.

Many people believe that anything you think will work, will. Sometimes they stop merely by taking your mind off them and thinking about something else.

More than one grandmother has recommended breathing into a paper bag. When we do, we breathe in carbon dioxide and it changes our breathing.

The fact is, the hiccups usually stop in a short time no matter what we do. Maybe doing nothing is as good a cure as any.

For some people hiccups continue for hours or days. This time can wear a person out and make him very sick. They usually need to contact a doctor and get professional attention.

The time we notice breath the most is when it acts strangely. That is the way it should be. God wants us to enjoy life and let him take care of our everyday breathing.

"For the soul of every living thing is in the hand of God, and the breath of all mankind" (Job 12:10).

1. What do you do when you get the hiccups?
2. How do hiccups begin?
3. How did mankind first get breath? (Gen. 2:7)

Thank you for being a watching God. What I cannot control, you keep in your hands.

twenty-nine

A Bladder Problem

School children who wet the bed face a confusing problem. Most of them hate it and would give anything to stop. What makes them do it when it is so embarrassing?

The first thing they need to know is that many children their age wet the bed. Some continue into the high grades. They are not strange or odd. Children with short fingers are not bad. Those with curly hair are not oddballs. We are all made differently and our bodies change at their own rate.

Some parents as they were growing up wet the bed for many years. Now they might not care to talk about it.

Our bladder holds the liquid called urine. It is the extra fluids our body does not need. When we are awake we can feel how full it is. If it becomes uncomfortable, we go to the bathroom and empty it.

Some children cannot feel this pressure at night. Therefore the bladder releases its urine without waking up the child. As they grow older the bladder will become stronger and awaken the person. Some merely will not hold much liquid and must empty during the night.

For some children there is no way to control this. He cannot say, "Today I have decided to become short." Neither can he make his bladder strong by hoping.

Embarrassing the child does not help. Discussing it every day only makes him feel badly. The only safe choices for the child are to let him grow out of it or take him to a doctor.

Not all children who wet the bed have a bladder problem. A few are too lazy to get up at night. They would rather sleep in a wet bed than walk to the bathroom. Hopefully this child will soon take more pride in himself. He will decide to become neat and take care of his clothing better.

Other children may be wetting the bed because they are lonely and want attention. Maybe he feels his parents don't spend enough time with him. This child needs to go to his parents and try to tell them how he feels. If he can't, he needs to talk to a school counselor or a minister.

The child who has a bladder problem must not give up. This cannot become an easy excuse for wetting. Avoid drinks before going to bed and always make the bathroom the last stop.

But they should not become discouraged. Someday it will stop.

God loves the child with poor eyesight. He loves the one whose feet are a little bigger. He loves the child who doesn't get A's. And he loves the child who wets the bed. And your parents love you, too.

"We know how much God loves us because we have felt his love and because we believe him when he tells us that he loves us dearly" (1 John 4:16).

1. Do you know anyone your age who wets the bed? Do children tease him?
2. What is a bladder?
3. What might God want someone who wets the bed to do?

Thank God for loving me just the way I am.

thirty

Is Bald Beautiful?

Every once in a while we see a bumper sticker on a car which claims "Bald is beautiful." Most of us have relatives who are either bald or partly bald. Some scientists tell us the men in the future may be all bald. Certainly we have less need for hair than we used to.

Actually all adults are losing hair. Most of us drop off 75 hairs each day. However, the majority of these grow back.

Often baldness is inherited. That means if our dad is bald, the chances are his son may become the same way. Baldness seems to affect men much more than women.

All the rumors about baldness coming from too much sun, too little sun or wearing a hat seem wrong. Most experts feel few things can help stop baldness. It appears to come from glands, hormones and heredity.

The person who keeps his body healthy may have a better chance of good hair. But he may be in the greatest condition and still have his hair fall like leaves.

The same problem centers around gray hair. Often parents want to know how to keep it from coming. There is probably only one way. They will have to color it. Gray seems to come when it wants to and there is little we can do to stop it.

Hair color comes from "pigment." As we grow older many of us lose our pigment and hair turns gray or even white. Often it begins around the temples. Sometimes it starts at an early age. Most of us will be like our parents.

There are a lot of jokes about baldness but many consider it a handsome sight. Some will shave their heads even though they have perfectly healthy hair. Others will add gray to go with their natural color. Still others will buy toupes or wigs to hide their scalp. There are some who hate baldness so much they have operations to move hair from their body and place it on their head.

During Bible days some men would shave their heads —especially if they were sad, when their country lost a war or were captured as slaves.

God didn't like to see people shave their heads because of sadness. Maybe they would want to shave it because they thought it looked good. But don't shave your head because you are sad.

All of us are hurt sometimes. A pet dies, a parent goes to the hospital, a good trip gets called off. Everyone has to be sad, but only for a while. God then wants to cheer us up and help us go on with life.

"Since you are the people of God, never cut yourselves [as the heathen do when they worship their idols] nor shave the front halves of your heads for funerals" (Deut. 14:1).

1. What causes gray hair?
2. What causes baldness?
3. Why did nonbelievers shave their heads?
   How does Christ help us during sadness?

Thanks for caring about our happiness and teaching us to keep our sadness short.

thirty-one

Bushy Beards

There used to be a famous softball team called "The House of David." People talked about them for two reasons: They were excellent ball players and each team member had a full, long beard.

Sometimes beards are very popular and a lot of people grow them. At other times most men shave and keep the hair off their chins. Practically all men have to shave or they start to get a beard.

The average man has 30,000 hairs waiting to grow out of his chin and jaws. In one year he can produce a beard 5 1/2 inches long. His face hair grows a little faster than that on top of his head.

Teenage males will notice hair begin to show on their cheeks and chin. Both boys and girls start to get hair on their bodies at this age. It appears under the arms and between the legs.

The purpose of hair on the body isn't easy to explain. In the days when people lived outside most of the time, hair on the face and chest probably helped keep the person warm. It may also have stopped a few bugs from getting into the mouth.

Most women cannot grow a beard or get hair on their chest. Some females get a noticeable amount of hair on the chin or above the lip but not to the extent of men.

Many of us used to think girl hair grew faster on the head than boy hair. We probably thought that because women wore their hair longer than men. The fact is that it grows at the same speed for both of them—about 5 inches each year.

Animals have more and better hair than humans. When

they get cold their hair stands on end to help protect the body. People have tiny hairs on their arms and legs which try to do the same thing. If we are chilly or sometimes afraid, our hair tries to stand up. Since we have so little all we usually get is "goose pimples."

Everyone seems concerned about having nice looking hair. Many adults spend a large amount of money to keep it looking nice.

The best way to have healthy looking hair is to have a healthy body. When we neglect ourselves it will show up in our hair.

Scientists are experimenting with hair to measure a person's health. Often we get blood tests to see if there are any diseases in our body. They believe some illnesses can be seen in our hair and beard.

If we suffer from too much lead in our body it will show up in our hair. Detectives can check into this. If a gentleman has been poisoned by arsenic, a scientist could find it in the dead man's beard. Traces of diseases may show up in our hair before the rest of our body knows about it.

During the old days of Israel most men prided themselves in their handsome beards. When they were sad they might pluck hairs out of it as Ezra did.

King David once sent special ambassadors to the funeral of a friend. The dead man's son didn't trust them and called them spies. He arrested them and shaved off half their beards.

"When David heard what happened he told them to stay at Jericho until their beards grew out; for the men were very embarrassed over their appearance" (2 Sam. 10:5).

1. How much does the average beard grow each year?
2. What are scientists learning from hair?
3. Why were the ambassadors' beards shaved half off?

Help us take care of our body so it will all be healthy.

thirty-two

Why We Get Thirsty

Many adults are carrying 100 pounds of water. We can't see it, but that is how much water there is in our bodies. If a child weighs 80 pounds, 50 of those are probably liquid.

Water fills every part of our body, even the teeth. Without it nothing in us could live. This is why people all over the world are always in need of water.

Most people drink 1 1/2 quarts of liquid every day. This may include a glass of milk, a bottle of soda, orange juice or water. We take in another quart from what we eat. There is liquid in mashed potatoes, green beans and peaches. All together we take in one-half gallon of water daily.

Water is more important to us than food. Pretend someone would lie completely still and drink nothing. It is possible to go 8 to 12 days without water. However, healthy, active people need their daily supply. When we play on a hot summer day our bodies lose water rapidly and we need to drink more.

It is easy to know when we are thirsty. Our mouth and throat usually get dry and sometimes we even feel weak. But this is not really the reason. Our body calls for water when our blood needs it.

When blood lacks water it starts to get thick and slow. At this time our blood sends a message to the brain and it calls for water. That is when we get thirsty.

After we drink some water our body tells us when to stop. We aren't sure where this message comes from. If we don't listen to this signal we will get sick from drinking too much.

Once a man drank 20 quarts in one day. That equals 40 pounds. His tough body was able to handle the flood, but it had to work hard to keep up.

After we eat peanuts or other salty foods, we notice ourselves getting thirsty. This isn't just in our minds. The human body has a careful mixture of salt and water. If we have too much salt, we need to add more water.

The same thing is true if we sweat too much. The water carries off salt from our system. Therefore, many athletes take salt pills to keep their body salt and water in balance.

People are now able to work in the hot deserts because they have learned to take care of their bodies. They use large amounts of water and salt as well as shade.

Sometimes Jesus Christ lived in the hot, desert parts of his nation. Water was always important to the people. He taught us to be kind to each other. In his day one of the nicest things anyone could do was to share his water.

"And if, as my representatives, you give even a cup of cold water to a little child, you will surely be rewarded" (Matt. 10:42).

1. How much of your body is water?
2. Why do peanuts make us thirsty?
3. What did Jesus Christ teach us about thirsty people? How else can we help people with needs?

We want to be a friend to people who are lonely. This is what Jesus Christ taught us to do.

thirty-three

Air Bags

Boys and girls enjoy running and playing games. Not only is it fun but it's healthy. We have strong bodies which can use more exercise than most of us give them.

Children have to spend hours every day sitting at a school desk. However, their lungs are made to take eight times as much air as they get sitting down. When we grow up our lungs really suffer. If we have an office job and don't play tennis or go bicycle riding, our lungs can't get the movement they need.

Most of us probably think of our lungs as being two big balloons. They do get large and small as we breathe but their makeup is closer to sponges. They are not empty but are filled with tissue which takes in air. There are about 300 million of these small air bags in the average man.

If these tiny tissues were stretched, our lungs would furnish half a tennis court—at least 600 square feet.

Our lungs take in the air our body needs to survive. But they have to have air in just a certain way. It must be moist and warm. This is where our miraculous body has to work together.

When we take in air our mouth and nose do an instant job of moisturizing it. In that same instant our blood vessels furnish the heat. Presto! Our lungs have their plate of air served up just right.

Once air is in the lungs two important jobs are done. First, oxygen is put into our bloodstream. It soaks through a very thin wall. Second, the lungs draw carbon dioxide out of our blood system. Therefore the air we exhale has more carbon dioxide than what we breathe in.

Our body burns the oxygen and gives us energy. But this is only one job. If our lungs did not carry the carbon dioxide out, it would poison our bodies.

It is easy to tell the difference between a child's lungs and an adult's. The child's are pink and fresh. The adult's will be gray from 40 years of pollution. The main function of the lungs is to give clean air to the body. Sometimes it seems as though we are trying to make their job tough.

We breathe in cigarette smoke, gas fumes and dust. After a while the lungs will not be able to carry clean air into the bloodstream. They will be so dirty they can deliver only dirty air.

"But you have not praised the God who gives you the breath of life and controls your destiny!" (Dan 5:23).

1. What is inside our lungs?
2. What do lungs do?
3. How can we praise God by keeping the air pure?

Thank God for plenty of clean air. Teach us to keep it pure so we can live as long as you want us to.

thirty-four

If No One Could Talk

It is hard to think of a world where people couldn't speak. Try to picture all the playgrounds filled with children but no sound. There would be no singing or stories read. No one could ever talk at the dinner table. They would only sit and point.

For one evening parents might think this was fun, but soon we would all miss it. People get much more out of life merely because they can talk.

In order for a human being to talk, three fairly simple movements take place. The lungs force air up toward the mouth. As it travels, air hits the vocal cords and causes them to vibrate making sound. However, the type of sound is dependent on our teeth, mouth, lips, tongue and jaws.

We do it so fast and easily we hardly notice these three areas in operation. We know just how to make a "g" or a "th" sound without even thinking about it.

If we want to make a loud noise we give it a large amount of air. If we decide to hum we release air and hold it back with our mouths.

Many children have high voices because their cords are still short. The air channel hasn't finished growing and they aren't as able to control the sounds they make.

As they grow older their voices will change. A person's voice will not depend on his size. A large adult may have a high pitch or a deep one.

We can see the location of the voice box on many men. The large lump on his neck, called the "Adam's Apple," is the bend where it rests. Among thin men it is particularly noticeable.

Among women the bulge is smaller and usually covered by a layer of soft fat. In most cases it can't be seen.

No two people have exactly the same sound. Their cords, mouth and teeth are all shaped differently.

God didn't give animals the same ability he gave humans. They can make sounds but not words in the same way we do. Some people who work with dolphins believe this animal may eventually communicate with humans. Maybe we will learn to speak its marine language or possibly it will adapt some of ours.

Speaking is a privilege and responsibility. Sometimes we talk and make someone feel happy and wanted. Other times we throw out a few words and really crush our neighbor. Our voice can be a welcome friend.

"Don't criticize and speak evil about each other, dear brothers" (James 4:11).

1. Name the three steps to talking.
2. Why do some voices sound high?
3. Why do we enjoy talking about other people?

My voice is such a good gift, I want to use it to make other people happy.

thirty-five

What Alcohol Does

Most American homes have some sort of alcoholic drink in the refrigerator. It may be beer, wine, whiskey or a number of other drinks. With so much around it would help to know something about it and its affects on the body.

How much a drink may disturb the mind or body depends on the amount of alcohol in it. Many beers have only 3% or 4%, though a few contain 15%. Wine usually has 7% to 23%. Strong drinks such as whiskey, brandy, rum and gin may have 50%. Therefore it takes more beer to affect a person than whiskey. How much a drink may bother someone depends on many things. The person's size, work, thinking, health and how much he eats all play a part.

Alcohol has two affects on the body and mind. In some ways it "picks up" the person. At first he may feel better, want to eat more and feel his pulse beating faster.

The same beverage is also a "slow down." Alcohol makes a person think slower. His body doesn't move as fast. He does things he would not do if he had been drinking milk or juice.

Because of these changes in his thinking and movements, it is dangerous for him to drive a car after a few drinks. His leg may move slowly to the brakes. He may think other cars are farther away than they are. One-half of all the deaths in car accidents are caused by someone drinking alcohol.

It is important to know that many people drink alcohol and never become drunk. We should also remember that many do get drunk and alcohol can be highly dangerous.

When people drink too much over a long period of time the body starts to get sick. Often they do not eat enough. Some people who "die a drunk" actually die from starvation because they have given up eating.

The most famous problem faced by heavy drinkers is the hardening of the liver. It is called cirrhosis and also happens to some nondrinkers. However, eight times as many heavy drinkers get the condition.

After drinking too much many people become sick. They vomit and have painful headaches. This is called a "hangover."

Those who drink in small amounts have no serious brain problems. However, those who drink too much for years often have some brain damage. If this happens because of too much alcohol or too few vitamins, we are not sure. These two usually go together.

Because of the lack of good food many who drink too much are often sick. They may catch pneumonia and other diseases easily and can't get rid of them well.

God has always been our friend and has warned us about things which hurt us. Too much alcohol can only cause trouble for our bodies.

"Don't drink too much wine, for many evils lie along that path; be filled instead with the Holy Spirit, and controlled by him" (Eph. 5:18).

1. What two things does alcohol do to the drinker?
2. How does too much affect our liver?
3. What do your parents think about drinking alcohol?

While we are young, teach us to be careful of things which could hurt us.

thirty-six

Sneezing Easy

Watch out when someone sneezes. Unless he covers his mouth and nose, germs are about to fly about three feet. These germs are able to carry diseases and make people sick.

Most of the spray coming from a sneeze races out of the mouth. But quite a few germs also escape from the nose.

One scientist has done research and discovered that 19,000 droplets fly through the air when one sneezes. A half hour afterwards hundreds of these wet germs are still alive in the room.

Every day we breathe in thousands of tiny pieces of dust, and these don't make us sneeze. Our nose is made well to handle whatever comes. However, sometimes an extra large speck of pollen or dust gets in. It is too big to be taken in or covered. A fast message comes from our brain, "Get rid of that."

Quickly we draw in a supply of air. Then in a sharp, strong blow air rushes up the nose and out the mouth. The farther back the dust is, the more wind we make and the noisier the sneeze. If someone sneezes over and over again, he is having trouble cleaning the intruders out.

Because of the thousands of germs present, it is always best to have a handkerchief close. We don't want anyone else to become ill because of our sneeze.

Coughing is caused in the same way. It often happens when we are eating. Food starts to go down, and because of size, shape or direction, it irritates our throat.

In this case we have no control over the cough. A message goes directly to the midbrain saying, "This feels strange." A command comes back just as fast saying, "Get rid of it." We then start to cough whether we want to or not. Sometimes it is embarrassing. Our face turns red and our eyes may fill with tears. We will keep coughing until the blockage is cleared up.

Not all coughing comes from something in the throat. A dryness from a cold will make the throat itch and we will cough to try and clear it. Smoke and too much singing will cause it.

Some people cough out of habit. They may be embarrassed or nervous. They are forcing themselves to cough.

Many people can cough because someone else does. If they are sitting at a table when another person keeps coughing, it will not be long before he does it also.

Colds are often spread from hand to mouth. The person coughs into his hand so he will not spread germs. Then he shakes hands with someone he meets. That person in turn touches his own mouth and starts his own cold.

In some societies people wear white mouth and nose masks during cold seasons. They hope to prevent spreading or catching germs.

God gave us a "smart" body. It tries hard to throw off strange things which don't belong in us. Otherwise man would be weak and couldn't live very long.

Elisha brought a child back from death. The first thing the lad did was sneeze.

"Then the prophet went down and walked back and forth in the house a few times; returning upstairs, he stretched himself again upon the child. This time the little boy sneezed seven times and opened his eyes" (2 Kings 4:35).

1. How many droplets escape when we sneeze?
2. Why should we carry a handkerchief?
3. Can we forget to sneeze?

God gave us special force to throw out strangers trying to sneak into our bodies.

thirty-seven

Funny Bone

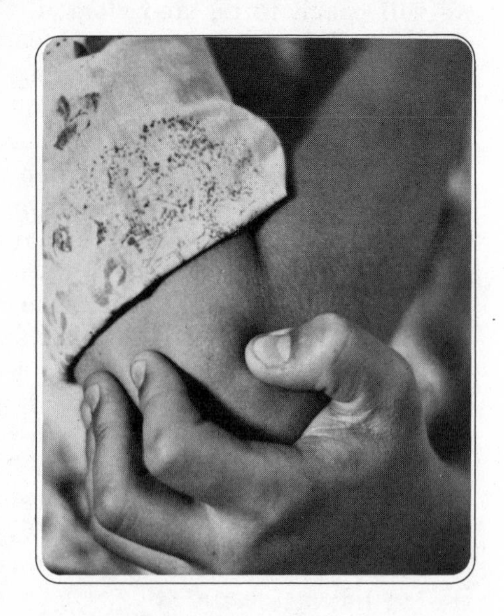

It's a terrible feeling. We move our arm back quickly and hit our elbow on the chair. It is a tingly feeling and yet it hurts at the same time.

We call it a "funny" or "crazy" bone, but that isn't the real cause. There is a nerve which runs from the shoulder down our arm. When it passes the elbow it is on the outside of the bone and close to the surface. It is called the ulnar nerve. If it is hit right, this nerve sends out the strangest feeling.

Not all of our arm causes this much notice. Most of it goes on doing its job day after day. The upper arm is the largest bone in the top half of our body. It is one bone which gets large at each end.

The forearm runs from our elbow to the wrist. It has two bones which are lined up side by side. These double bones are important to our hand. They allow the wrist to twist and hand to turn around. The two are called the ulna and the radius.

Hold your palm up and look at it for a moment. Now the two forearm bones are in the right place. Turn your hand over and look at the back of it. The two bones are now twisted with one laying over the other.

The bones in arms often break, but not because they are weak. Bones are some of the strongest materials on earth. Our leg bones will hold thirty times the weight they are now carrying. Some bones will hold a small truck without breaking.

Iron could not withstand the activities a body goes through daily. The weight and movement which bones take are too much for almost any building material.

Bones have their limit. However, they withstand constant action. Practically any machine would fall apart under years of continuous use.

When God made the elbow he created a simple pulley. The muscles running through the arm work like a rope. Take a cup in your hand and see how it works.

Lay your arm flat with the cup facing you. Now draw it up toward your head. As you lift, the muscle in the upper arm pulls. It tightens the muscles in your forearm and hand and draws it up. To lower the cup merely relax the rope (muscle) and the bridge will lower.

It is called a "funny" or "crazy" bone, but it is one of the most amazing machines in the world.

Some days we need to stop and be thankful. Not everyone has strong bones. They can't run or throw a ball. Others of us enjoy our bones every day. Once in a while we should tell God we appreciate it.

"You gave me skin and flesh and knit together bones and sinews. You gave me life and were so kind and loving to me, and I was preserved by your care" (Job 10:11, 12).

1. What makes the elbow a "funny bone"?
2. When are the forearm bones straight?
3. Give three reasons why you are thankful.

***Healthy bones are a gift.
We have a lot to be happy about.
Over 200 reasons just in
our body.***

thirty-eight

An Old Illness

The body is usually a strong, healthy machine. It works so well that man has not been able to build anything like it. But sometimes a disease attacks the body and it can't get rid of it.

No disease has been more dreaded than horrible leprosy. The name sounds ugly to most of us. It is a serious disease, but many things said about it are not true.

One hundred years ago Doctor Hanson studied leprosy and today it is called Hanson's disease. He found leprosy came from little plants (fungus) growing inside the skin. They would make lumps on the body which at first could not be touched without pain.

Later these lumps became hard. A pin prick would not hurt one. Even a lighted cigarette could not be felt. Doctors would then remove one of these lumps to see if these plants were inside. If they were, the person probably had leprosy.

After the disease had been around for a while, the victim might start to lose parts of his body. Usually it took the end parts like fingers, ears or the nose.

In most places special villages were set aside for people with leprosy. They were not allowed to mix with anyone healthy. Some lepers died from old age. In others, the leprosy spread until it killed them.

Today there is a cure for the disease. There are treatments with drugs which kill the plants (fungus). But millions of people still have leprosy in Asia, Africa and the Philippines. Some missionary doctors say the biggest reason leprosy has not been wiped out is the lack of money to buy the drugs.

There are people with leprosy in the United States. In Carville, Louisiana, there is a National Leprosarium which has around 50 patients each year. New York City reports nearly 25 cases yearly. Most of these victims were born in other countries.

When Jesus walked in Israel there were many lepers around. Because the disease was so bad, people would never touch a leper. Jesus knew better. He knew the disease was hard to catch. People did not get it by just touching a leper. Nor did children get it from their parents. Missionaries and doctors work for years with lepers and don't get it.

When Christ passed through Galilee, a leper knelt in front of him and asked to be healed. Jesus was moved by the sight of the man. He reached out, touched him and the leper was healed.

Christ did two things: He healed a man and he touched a leper. Hopefully other people would help them instead of being so afraid.

"And Jesus, moved with pity, touched him and said, 'I want to! Be healed!' Immediately the leprosy was gone—the man was healed!" (Mark 1:41, 42).

1. What causes leprosy?
2. What is leprosy's other name?
3. How did Christ help?

Christ has shown us to love unlovely people. Help us to "touch" those whom no one else wants to be around.

thirty-nine

Black Cats and Warts

Could you get rid of a wart by burying a dead cat at midnight? Will they disappear if you put moth wings under your pillow? Both sound like fairy tales, don't they? But before we laugh too hard, we should know there might be some truth to these stories.

Science doesn't know as much about warts as they would like. But it looks as if some of these little lumps on our skin fall off because we think they will. It's called therapeutic suggestion. In this case, by believing it will happen it just may.

Children seem to get warts more often than adults. They are most likely to collect them on the back of their hands and fingers.

Warts are caused by tiny living viruses which can be seen only under microscopes. Somehow they grow and can be terribly uncomfortable. Often people get them on the soles of their feet which can make walking painful.

One woman has to visit a doctor every two weeks to have warts cut out.

Chemicals have been successful in removing many warts. However, because of the many stories which have spread about them, warts are often thought to be magical.

These bumps also grow frequently on middle-aged people. They may get yellow spots on their eyelids. These tiny warts are only the size of a pin head but they can grow larger. This may be the result of too much of a certain food in their diets. It can also mean their blood pressure is getting too high.

Some warts spread easily. For the best results the person may want to visit a pharmacist or doctor. Many young people feel particularly ugly with them and may go to extremes in trying to hide them. Help is not difficult to get.

Once in a while we wonder what kind of body Jesus Christ must have had. Did he have any scars, pimples or acne as a child? Could he have had any warts?

We can't be sure about most of these questions. Everyone must decide for himself.

We are sure his life was perfect. He lived with brothers, sisters and neighbors and still managed to not sin. That was a difficult thing to do. The only reason Jesus could was because he was the Son of God. Now he wants to teach us to stay away from sin also.

"But he paid for you with the precious lifeblood of Christ, the sinless, spotless Lamb of God" (1 Pet. 1:19).

1. What causes warts?
2. What cures warts?
3. How was Jesus spotless?

God has shown us a perfect life in His Son, Jesus Christ.

forty

35-Mile Filter

Drinking is dangerous if water is dirty. That is why your community spends a good deal of money to clean the water. The human body would be in trouble if we didn't have something to keep our blood clean. That is why God gave us two kidneys.

These bean-shaped organs are the major filter plants for our blood system. Every day 1,700 quarts of liquid travels through these "nets," and they must do the job quickly.

The kidneys remove the waste material from our blood and turn it into urine. The urine travels down to our bladder and waits. Several times a day we empty it out. About one and one-half quarts are passed each day.

The filtering kidneys aren't large, no bigger than a doubled fist. But it has so many small veins (capillaries) that if all of them were laid end to end, they would stretch for 35 miles.

In some parts of the forest stream water is pure. It has flowed for miles over sand and small rocks. The blood in our kidneys is filtered the same way. It is pushed

through the tissues in our kidneys as it filters. There are no special chemicals in the process. The blood is merely filtered as we would put dirty water through a cloth.

A major job for the kidneys is to control our use of salt. A body needs salt, but in correct amounts. Each day around 2 1/2 pounds of salt passes through and some is removed. If the kidneys allow too much salt to remain in our body, our tissues begin to hold too much water. However, not enough salt may make us sick, especially on hot days.

We cannot live without kidneys. If they stopped working altogether, it used to mean certain death. However, today there are two famous ways to replace them: One is the artificial kidney. It is a machine which can be hooked to a person's body. The patient's blood passes through the (dialysis) machine and is cleansed just like a real kidney.

The user can live a fairly normal life. He merely needs to visit the hospital and be cleaned out once in a while.

Another way is kidney transplants. In recent years kidney transplants have become common. People have agreed to give their kidneys when they die. Therefore someone whose kidneys have stopped working may get new ones.

A body needs to remain clean if it is to stay healthy. A person's life also has to be clean if his mind and soul are to be well. Therefore Jesus Christ died to cleanse us of our sins. We don't have to let selfishness and meanness control our lives.

"But if we are living in the light of God's presence, just as Christ does, then we have wonderful fellowship and joy with each other, and the blood of Jesus his Son cleanses us from every sin" (1 John 1:7).

1. Describe the kidney's job.
2. What does a dialysis machine do?
3. How can Jesus help us when we sin or do something wrong?

Sometimes our lives get dirty. We do what is wrong and we know it. Please keep us clean by forgiving and helping us.

forty-one

Picking Up Cars

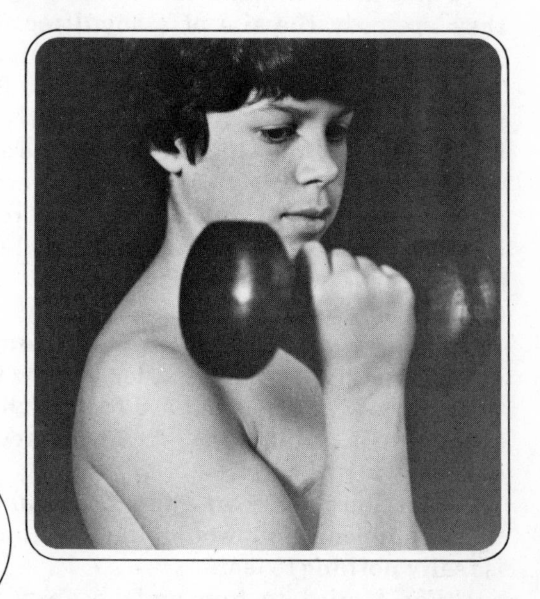

Have you ever heard a story of a man who picked up a gigantic log? Usually he couldn't lift something that big. But maybe his son was trapped under it. The man now found himself stronger than he had ever been before.

Some stories like this are definitely true. Our body is made to give us extra strength in an emergency. A woman might pick up one end of a car. A young boy might run faster than he ever had before.

There are two "emergency" glands which sit on top of our kidneys. They are called adrenal glands. If you get into trouble, these little sacks help out by sending extra sugar into our blood. This gives us an added dose of energy like a booster shot.

What is added to our body is called adrenalin. It comes immediately without having to be asked.

Not only does adrenalin affect our blood, but it also helps our muscles. We now have more strength than we normally have while working. Our body knows when we are afraid and sends a new supply of energy.

When we see a young boy running away from an older one, we know what is happening in his body. The smaller boy's adrenal gland is pumping sugar to get him going.

These important glands are not large. In a grown man they are only the size of a small toe. If this gland was destroyed we would not be able to live.

The adrenal gland has two parts. The outer section is called the cortex. The inner part is the medulla. It is the inner section which pumps up the adrenalin.

For some people this gland is overactive. It is regularly supplying too much energy, more than the person can use. With others the adrenal gland is too slow. People who are overworked or lack enough food may have a sluggish adrenalin flow.

When we are afraid we have two quick forms of help: our body starts working harder to help us and God comes to our aid. Many people have been frightened and started to talk to God about it. At once they felt calmer and safe knowing he cared.

Sometimes we worry too much about things which never happen. God would like to quiet us when there is really nothing to fear.

"For I cried to him and he answered me! He freed me from all my fears" (Ps. 34:4).

1. How does an adrenal gland help me?
2. Describe the parts of this gland.
3. Have you ever been afraid and talked to God? What happened?

Jesus told us to not be afraid. Help me to trust Him as a close friend.

forty-two

Rattlers

The human body is tough. It will do amazing things and fight off many illnesses. However, some injuries need help to overcome, and poisonous snake bite is one of them.

Most snakes in North America are not harmful. They are more afraid of us than we are of them. Unless they are trapped, almost all snakes will quickly crawl away from people.

Nonpoisonous snakes will bite but little harm comes from them. Their bites should be cleansed with soap and water and a doctor should be contacted.

Normally there are four types of poisonous snakes in our part of the world: rattlers, copperheads, moccasins and the seldom seen coral snake. Because of education and good medical aids not many people die in the United States from snake bite.

Most poisonous snake bites leave two holes which look like needle marks. It could be one hole if it bit from an angle. There might be four to eight marks if the snake quickly made several bites.

These punctures come from the fangs in a snake's mouth. The two long teeth have tiny needles in them which shoot out when it bites. The poison is carried in these needles.

In most cases a poisonous snake bite is highly painful. It also causes swelling around the mark. This may make it difficult to see the needle holes, and the marks are not always easy to make out. A few poisonous bites cause no pain. The only safe thing to do is see a doctor immediately. In areas where dangerous snakes are found, medicine is available quickly.

The only safe step is to get antivenom from a doctor's needle. It allows the body to do the job it can't do on its own.

It is important that the victim not move around much. Motion will cause the poison to spread rapidly.

There used to be a story that whiskey should be given to the victim. Doctors now tell us not to use alcohol. They also tell us not to suck on a bite if there are cuts or sores in our mouth. It is possible for us to get the poison.

Our bodies fight off many poisons and diseases. However, they have their limit. Any snake bite should send us to get help at once.

If we don't understand our bodies, we can get into serious trouble. Some people act as if they can eat or drink anything and it won't hurt them. After all, our body is young and tough. But there is a limit and we are foolish to go too far.

"Don't let the sparkle and the smooth taste of strong wine deceive you. For in the end it bites like a poisonous serpent; it stings like an adder" (Prov. 23:31, 32).

1. What poisonous snakes live in North America?
2. How can you tell if some snake bites are poisonous?
3. Can you name some ways in which we hurt our body by what we eat and drink?

Lord, teach us to be careful about hurting our bodies.

forty-four

Bloody Noses

Alan hadn't been wrestling with his brother long before his nose felt funny. He recognized the feeling right away as he had often before. His nose was bleeding and he knew he better not get blood on the floor.

This boy's problem is common and raises two interesting questions. What made his nose bleed? And why did it finally stop?

Often we think of our bloodstream as a few veins in different parts of the body. The fact is that the bloodstream in the average person is almost 100,000 miles long. It sounds too much to believe, but it is a huge system.

The blood is continuously moving through the body and being pumped by the heart. If any part of the body is cut or the skin is broken, blood starts to escape through the wound.

A regular rescue squad lives in our blood system. Its members do not cruise around because they would start to "gum up" the streams. But when the skin is torn or punctured they come running faster than a fire engine down Main Street.

First the light equipment called "platelets." They are little flat tissues to form a quick bandage.

Second, fibrin rounds the corner with wheels squealing. This is the heavy "glue" which will make a strong wall. It only takes a few seconds to show up.

These two have done the emergency work. For something as simple as a nose bleed they should begin to close the wound in a short time.

However, our blood isn't content to lock the door. A third member of the squad now moves in. Its job is to look for enemy germs and knock them out.

Called antibodies, these specialists fight spies like flu virus, pollen, splinters and hoodlums such as bacteria. Normally the antibody division can handle most intruders. Yet once in a while a super villain like tetanus sneaks in and the family doctor needs to be called for an extra shot of antibody.

If all the necessary members do their jobs the blood will "clot." Some people don't have all the parts and they may bleed for a long time. When each member does its job the seal starts to harden and a scab forms. After a while the scab falls off and the skin is often as good as new.

Most of the things which happen to our bodies are no surprise. If we run fast our heart "pounds." If we hit a thumb with a hammer, there is no doubt it will hurt. When we yell at someone, we usually get yelled back at. Some things in life are what we call "dependable." We know what will happen.

Normally, if we treat our brother, sister, father or neighbor kindly, they are kind to us. If we hit someone, we usually get hit back.

"As the churning of cream yields butter, and a blow to the nose causes bleeding, so anger causes quarrels" (Prov. 30:33).

1. Describe one of the members of the "rescue squad."
2. How long is our blood system?
3. Does Christ ever want us to get angry?

We want people to say nice things about us. It makes us feel good. Lord, help us to take the first step by saying something nice about them.

forty-five

Human Air-Conditioning

The next time you start to sweat, stick out your tongue and taste a drop. There is usually an odd salty flavor to it. This is because our sweat carries about one half of one percent salt.

If we sweat too much, as athletes might, we need to take salt tablets. When too much salt leaves our body we could become sick. Otherwise, our sweat consists of pure water.

Practically everyone sweats and we never think much about it. But what would happen if people didn't perspire? Dogs don't sweat so they stick out their tongues and pant. Pigs have to roll in the mud to cool themselves. People don't do either because they have an excellent air-conditioning system.

Our body temperature normally stays around 98.6 degrees. If we work or play hard, we become warmer. When the sun beats down on a hot summer day, our body also becomes hot. To keep us from overheating, God put a special and automatic cooler in us.

When people get too hot, water rises out of the skin. It comes out through tiny holes we call pores. If we look carefully, we can see a few on our arm or hand. Behind these pores is a network of 2 million small glands. The glands are little pumps which push the water out.

However, merely sweating is not enough to cool the body. The real help comes when the water evaporates. We may be sweating a great deal and still feel hot. But as the water dries, our temperature will drop slightly.

For all their ability, sweat glands can help us only so much. If we are not careful, we will overheat and get sick. There is a point at which we need to just quit and be good to our working body.

As children get older they notice a strange odor which comes with sweat. Usually people under 12 don't have this problem. At about junior high age our sweat changes just a little bit. When the "adult" sweat mixes with our outer skin it causes this new odor. It is true that underarms have a worse smell than other parts of our body.

Because people live and work so closely many spend large amounts of money fighting odor. In most cases we are trying to supply a pleasant odor in place of an unpleasant one.

God expected people to work hard in life. Some with strong arms, and lifting heavy materials. Others behind desks thinking, studying, teaching or selling. He put a small air-conditioning system in us so we could work or play without tiring too easily.

"All your life you will sweat to master it, until your dying day. Then you will return to the ground from which you came. For you were made from the ground, and to the ground you will return" (Gen. 3:19).

1. How does sweating cool?
2. How do dogs cool off?
3. Why did God give us the system?

God gave us "air conditioning" so we could work and play to get the most out of life. Show us how we can do both but at the right times.

forty-six

The Sugar Factory

Many people can name the heart, lungs and stomach. They are popular body parts which do important jobs. But there is one part that keeps us alive, and many children have never heard of it. Without it you couldn't eat a candy bar or drink a bottle of pop. There was even a day when the person would soon die if this part quit working.

It is called the pancreas. If it doesn't work correctly, we get something called diabetes.

Many people, including children, have this problem. You probably know a relative or friend who has it.

The pancreas lives in the same neighborhood as the liver and kidneys. Its main job is to give sugar to our blood. We need it for energy. Without it we could not flick our finger.

If the pancreas does not change sugar into body energy, we are in double trouble. First, we will not have enough strength. Second, sugar will begin to pile up in our body. The sugar spills over into our kidneys and passes out of our body.

Before medicine was found to help this, people almost always died from it. The medicine used now is called insulin. It is taken with a needle. It does what the pancreas has stopped doing. The diabetic has to be careful what he eats or he will get sick.

In 1921 two Canadians, Frederick Banting and Charles Best, discovered insulin. Without this many more people would have died. Best was a 22-year-old medical student.

Candy bars and other sweets do not cause diabetes. Sometimes in small amounts they give extra energy. But too much candy does make the pancreas work overtime.

This work is given to an organ which weighs three ounces. It stretches five to six inches long.

The fuel which comes from sugar is called glucose. Just a certain amount should be circulating in the blood. For most of us this is around a sixth of an ounce. The pancreas has to be careful that we get the right amount—not too much or too little. When the pancreas doesn't work correctly most people have to cut down on their eating as well as to take medicine.

A carefully arranged body such as ours did not happen by accident. It is balanced correctly. Any sudden change and it may act strangely.

The person who made the human body is a great, intelligent God. He is worth knowing, worshipping and serving.

"Don't let the excitement of being young cause you to forget about your Creator. Honor him in your youth before the evil years come—when you'll no longer enjoy living" (Eccles. 12:1).

1. What does a pancreas do?
2. Who is Charles Best?
3. What does our body tell us about God?

Thanks for the knowledge of what is in our bodies. It makes us understand how great our God is.

forty-seven

Fast Feet

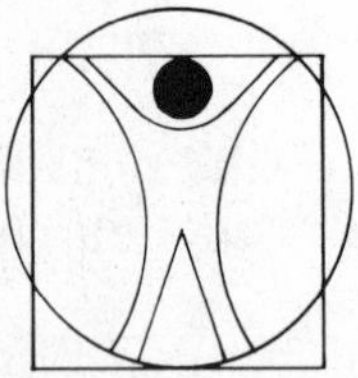

Why do feet curve? Most of us have an arch under our foot. When we walk, run or shift from one leg to the other, our feet go up and down like a sponge.

Those who don't have this arch have what is called "flat feet." They have trouble walking or running long distances. Their feet hurt easily because they don't have a sponge.

A look at the bottom of our foot will show us one arch, but actually there are three. Two of them stretch from heel to toe and the third goes across from left to right.

These springs are important to every active young person. They allow us to run, jump and change directions quickly.

The arches do more for the body than give us quick, comfortable feet. These sponges are necessary for the protection of our spine or backbone. Without arches each heavy step would shake our spine. After a time our backbone could hurt and even be damaged. Arches let us down easily.

Some habits will cause people to become flat-footed. Shoes which don't fit correctly, overweight and too much standing on hard ground may help destroy the arch. The muscles simply will no longer do their job.

The shape of the arch is not the only protection the foot and back receive. There is also a layer of healthy fat in the foot. It helps like a cushion in a chair or a mattress.

In some ways man really walks on four feet instead of two. Each foot has a heel and sole with the arch in between. The two parts give man added balance and movement. Without the sole, heel and arch, people would have trouble carrying such a large body on just two legs and feet.

There is a lot of fun in running—that's why we have races. There is a lot of help by running—that's when we hurry to get the cat out of the tree. There is a good deal of trouble in running because we think we can do something wrong and get away with it.

When Solomon thought of running, he remembered there were times when he should hurry to God. He was about to do something wrong. He was about to break a law. Any minute he was going to make a fool of himself. He hurried back to God. He talked to his Heavenly Father. He asked questions. He listened.

If he didn't hurry to God, he knew he would do something he would be sorry for.

"The Lord is a strong fortress. The godly run to him and are safe" (Prov. 18:10).

1. How is the arch related to the spine?
2. How are we "four-footed"?
3. Give an example of when the godly should run to God.

When the going gets tough we have someone to go to.

forty-eight

Where Did the Apple Go?

We took a big bite from an apple, and that's the last we saw of it. But what does it do in our body? Where does it go? How does it help?

To understand what happens to food in us, start by remembering four big steps. When all of them are working, our body factory does a great job.

The first step begins in our mouth. The minute it tastes food our mouth gets a liquid called saliva. It comes from three places around the tongue. Sometimes saliva flows just by talking about tasty food. This juice from our mouth starts to change the apple so our body can use it. By chewing the food many times we can use it better.

When we swallow our food it travels quickly to the second step in our stomach. Our supper doesn't merely sit around here. The stomach does almost the same thing our mouth did. Its powerful muscles "chew" on the food and make it even smaller. The stomach is also filled with juices like the saliva in our mouth. These juices continue to take the food apart.

Often the stomach takes about four hours to work on food. Then it is ready for the third step. Food moves into our two intestines.

The intestines are tube-like parts which are five or six times as long as our body. A five-foot-tall girl probably has a 25-foot intestine. As food moves through this tube it continues to give off valuable parts to the body. By the time our piece of apple has traveled through the intestine, its good food parts have been taken out. What is left will then pass out of the body.

The fourth important step takes place when the food enters the intestine. Our liver sends a juice called bile into our food. The bile gets rid of anything in the food which could make us sick. Sometimes we eat things too strong even for liver bile to protect. But most normal foods are cleaned up by bile. Presently if a person's liver stops working, he probably cannot live for long.

The digestive system has many more actions, but these four give us the easy outline. In four big steps a piece of apple passes through our body. On the way it gives the food we need to keep alive and helpful.

Jesus Christ knew the body was important. A person should be careful what he eats. But Jesus knew that what we think and say is more important. If we eat too many green apples, we might get sick for a while. But if we hate people and try to hurt them, we have done something far worse.

" 'Don't you understand?' Jesus asked him. 'Don't you see that anything you eat passes through the digestive tract and out again? But evil words come from an evil heart, and defile the man who says them' " (Matt. 15:16-18).

1. Describe the four steps to the digestive system.
2. What is bile?
3. Why do children and adults do mean things?

Thanks for the bile to keep our body clean. Help us to control our mind so it will stay clean too.

forty-nine

A Bent Back

The bone which runs up the middle of your back is called a spine. If we could see it, we would notice its two slight curves which make it look like an "S."

A spine is not just one long bone. It has 24 bones called vertebrae. They are stacked on each other like pancakes. In between each round bone is a soft disc. The disc works like a piece of rubber. If we bend or jump up and down, these rubber-like discs stop the bones from hitting against each other.

Our backbone is strong and can take a lot if it is treated correctly. It makes thousands of bends and movements every day. Children really give their spine a workout. This bone may hold a hundred pounds in sudden jerks and twists each day.

If a person works on his feet all day, the "rubber" discs start to shrink. When he goes home at night he may be one inch shorter than usual. After a good night's rest, his spine will go back to its regular size.

Most of us probably don't give our back much thought. So far it has always done its job. But a disc can be pushed out of place. This can be painful and need special attention.

The spine is not only important as a bone but it also helps our brain. Special lines run from all over our body up to the brain. These allow our brain to tell the body what to do.

These "telephone" lines travel through our spine to the back of our neck. There they give and receive information for the body. These nerves or "telephone" lines are protected in our backbone.

As some people grow older, their spine begins to bend forward. The same thing happens if we do not sit up straight. Our backbone becomes more curved than it would normally be.

Sooner or later most people complain of some type of back pain. Many times this is a result of not sitting correctly and taking care of our spine. Some people must wear special back braces to make their backbone straight.

The Bible tells of a lady who had serious back trouble. For 18 years she was bent almost all the way over. It must have been a real effort to walk at all.

Jesus saw her while he was teaching in the synagogue. Christ cared about her and healed her back. The first thing she did when she stood straight was to praise God.

"He touched her, and instantly she could stand straight. How she praised and thanked God!" (Luke 13:13).

1. Describe the job of a disc.
2. How do backs become bent?
3. Why did Jesus heal the woman?

Jesus healed bodies because
He cared for people.
Help us take care of ours
to be healthy for you.

fifty

The Pimple Problem

Between the ages of 12 and 25 most people get pimples on the face. This is called acne. Many young people worry about it, but most can do little about it except wait until they go away.

Doctors do not always agree on what causes acne. The biggest reason is just growing up. Around the age of 12 we begin to change. We get more hair on our bodies and start to become adults. This is called puberty.

During this change our body makes more hormones go into our blood. These extra hormones make pimples rise on our face. It is a change which happens to almost everyone.

Sometimes young people with acne are teased a great deal. Their friends even make up stories about what causes these pimples. The best thing may be to just laugh and not take them seriously. There is a physical reason for acne.

We have often heard that pimples are caused by too many sweets and candy bars. There may be some truth to this. Certain foods may make them worse. Chocolate, nuts, peanut butter, seafood and soft drinks are a few of the problem foods. A good selection of the basic foods such as fruits, vegetables, milk, cheese, meat, bread and cereal are probably more helpful.

It is always important to keep our face clean, but especially when acne comes. Some doctors will suggest as long as 15 minutes a day with a warm cloth on the face. Any picking at the pimples should be done carefully. Marks can be left on the face.

If anything seems unusual about a person's acne, a doctor certainly should be contacted. Since we are all different we need individual care.

During this time it is also important to keep hair extra clean. Twice weekly we should shampoo with hard scrubbing followed by a good brushing.

God wants all of us as healthy as possible: the parts inside our body which we don't see, our skin, face and hair. He also wants us to have a healthy mind.

Some young people let pimples bother them too much. It is important, and no one wants them. However, in most cases they go away. Too strict a diet, special kits bought through the mail and even avoiding people may be taking this too seriously. Most of us pass through it very well. If not, then we need to take more serious steps.

When acne comes, try to relax. It means you are growing. Another big step in life.

"Dear friend, I am praying that all is well with you and that your body is as healthy as I know your soul is" (3 John 2).

1. How many people do you know who have acne pimples?
2. What probably causes them?
3. How can God help me go through this part of life?

Help me accept pimples as a normal part of a healthy life. Don't let them bother me too much.

fifty-one

The Control Tower

Most of us have seen a picture of a skull and crossed bones. They appear on some bottles which have poison. Pirates used to paint them on their ships and flags.

If a skull is scary looking, it is also one of the most interesting parts of man. The bones of a skull are easy to understand. All totaled, our entire head consists of 22 bones and only one of these moves.

Let's begin by dividing our head in two parts. The upper half is called cranium and has just 8 bones. The lower part is our face and has the remaining 14.

The face has a number of holes to help us live in our world. There is an open space for our eyes, ears and nose. There is also a horseshoe shaped bone we call a lower jaw. It moves up and down allowing us to eat, sing, talk and even yell. The jaw fits into the sides of our cranium by tiny hinges on the sides of our face.

The upper half of our head houses our brain and serves as the control tower. You have probably seen a tower at an airport. People in this room tell all the planes when to come in and take off. Our cranium can rapidly tell a part of our body what to do and it will obey.

In the back of our skull there is an opening at the base. Through this hole runs our backbone or spine. Our spine holds nerves like telephone wires. They carry messages between our brain and all over the body. There are 31 pairs called spinal nerves.

It is a fast telephone system. If my finger touches a fire a message runs quickly to my brain and a call comes back telling my hand to "move it." It doesn't take my finger long to obey.

For some reason these lines are connected in an odd way. The cords which reach the left side of the brain control the right side of the body. Those on the right side control the left. Doctors aren't sure why they cross over.

It is a good thing we have only one head. Can you imagine two brains giving orders? Suppose it starts to rain. One brain would say, "Duck under that tree." The other cranium would say, "No! Run for home." One leg might start running and the other leg might try to hide.

God gave us one good head to control a busy body.

The Bible compares Christians with the human body. We are many parts. Some of us are short, some fat, bald, quick and slow. With all our different parts we need one, good head. That head is Jesus Christ. We can send messages to him and he in turn helps run our lives. Life is confusing with so many people telling us what to do. Thank God that our Christian life has only one head.

"He is the Head of the body made up of his people—that is, his church—which he began; and he is the Leader of all those who arise from the dead" (Col. 1:18).

1. Describe our head bones.
2. How does our "telephone" system work?
3. How can Jesus be my control tower?

Jesus Christ is in the control tower. He wants to help me by directing my life.

fifty-two

Nature's Best Tool

If you need a drink but don't have a cup, just reach out your hand. How about a hook to carry a bucket? A hammer to pound? Some tweezers to pick weeds? How about a brush to push your hair back? Or a pointer or a shooter for marbles? A shovel to dig dirt, a paddle for swimming, or a crane to pick up rocks and boards.

The hand is one of the best tools God has made. It is better than a metal or wooden tool. The human hand can also feel and help the person make decisions. Is this dirt too wet? Is that apple too soft?

Our fingertips can feel things better than any other part of our hand. Only our lips, tongue and tip of the nose can feel better.

The hand has 27 bones. Eight are in the wrist and the other 19 in the fingers.

When God made the hand he had some problems to solve. How can a hand be made which is both movable and strong? The muscles would have to be so large our fingers would be fat and hard to close. The answer was found in using a large number of small muscles. He put 15 muscles in our big forearm. But he placed 35 small ones in our hand. That's why our hand can do small jobs and still have good strength.

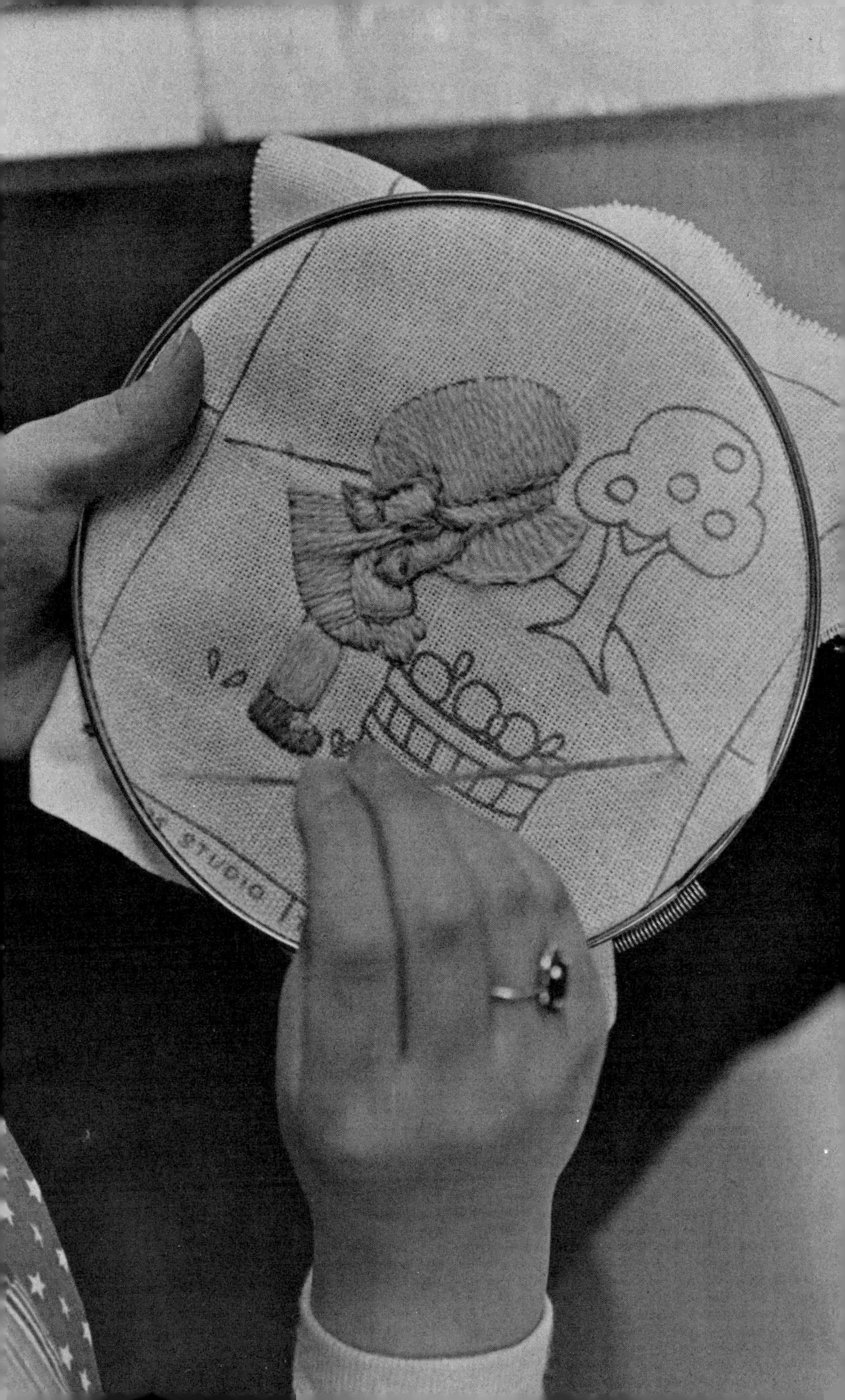
STUDIO

Each finger has an easy-to-remember name. Begin with the thumb. It is stronger than the rest of the fingers. If we have to push something hard, we often use it. The "thumbtack" gets its name because it takes a thumb to push it.

The next finger is called the "index" or pointer. It makes a good stick or lets us count papers quickly. The third finger is called the middle and is slightly longer than the others.

The fourth finger is called the "ring." In North America many married people wear their engagement and wedding rings here.

The last one is called the "least" or "little" finger. Sometimes it is named our "little pinky."

Each one of our fingers is a different size. This allows us to do difficult jobs more easily.

Because the hand is made so well, it can do things which need both strength and gentleness. A person may use a hammer in the morning to drive nails. In the evening that same hand may paint a picture. It might hold a baseball bat one day and a careful doctor's tool the next.

Hands do a great deal of good for man. But the same ones are able to do a lot of evil. It all depends on how we want to control them. Sometimes hands get us into trouble simply because we have not found anything good for them to do.

"Idle hands are the devil's workshop; idle lips are the mouthpiece" (Prov. 16:27).

1. What is different about hand muscles?
2. Name each finger.
3. Give examples of idle hands getting people into trouble.

Please help us to control our hands. They are amazing tools but they need to be told what to do.

he gave